T0215717

# CAMBRIDGE LIBRARY COLLECTION

*Books of enduring scholarly value*

## Life Sciences

Until the nineteenth century, the various subjects now known as the life sciences were regarded either as arcane studies which had little impact on ordinary daily life, or as a genteel hobby for the leisured classes. The increasing academic rigour and systematisation brought to the study of botany, zoology and other disciplines, and their adoption in university curricula, are reflected in the books reissued in this series.

## Studies of Plant Life in Canada

Catharine Parr Traill (1802–99) was a writer, botanist and settler who emigrated from England to Canada with her husband in 1832. Both she and her sister, Susanna Moodie, became well known for their writing on settler life: Traill is also the author of *The Backwoods of Canada* and *The Canadian Settler's Guide*. This 1885 publication is the most comprehensive of her botanical works. Plants are grouped together by family and the book is divided into four sections: native flowers, flowering shrubs, forest trees and native ferns. Written to inspire the Canadian public to share her passion for the plant life of their country, the book has an engaging style where anecdotes and literary quotations appear alongside detailed descriptions and classification information. Traill's niece, Agnes Chamberlin, is the book's illustrator. A beautiful example of nineteenth-century popular botany, this book will appeal to anyone interested in the history of the subject.

Cambridge University Press has long been a pioneer in the reissuing of out-of-print titles from its own backlist, producing digital reprints of books that are still sought after by scholars and students but could not be reprinted economically using traditional technology. The Cambridge Library Collection extends this activity to a wider range of books which are still of importance to researchers and professionals, either for the source material they contain, or as landmarks in the history of their academic discipline.

Drawing from the world-renowned collections in the Cambridge University Library, and guided by the advice of experts in each subject area, Cambridge University Press is using state-of-the-art scanning machines in its own Printing House to capture the content of each book selected for inclusion. The files are processed to give a consistently clear, crisp image, and the books finished to the high quality standard for which the Press is recognised around the world. The latest print-on-demand technology ensures that the books will remain available indefinitely, and that orders for single or multiple copies can quickly be supplied.

The Cambridge Library Collection will bring back to life books of enduring scholarly value (including out-of-copyright works originally issued by other publishers) across a wide range of disciplines in the humanities and social sciences and in science and technology.

# Studies of
# Plant Life in Canada

*Or, Gleanings from Forest, Lake and Plain*

CATHERINE PARR STRICKLAND TRAILL
AGNES FITZGIBBON CHAMBERLIN

CAMBRIDGE
UNIVERSITY PRESS

CAMBRIDGE UNIVERSITY PRESS

Cambridge, New York, Melbourne, Madrid, Cape Town,
Singapore, São Paolo, Delhi, Tokyo, Mexico City

Published in the United States of America by Cambridge University Press, New York

www.cambridge.org
Information on this title: www.cambridge.org/9781108033756

© in this compilation Cambridge University Press 2011

This edition first published 1885
This digitally printed version 2011

ISBN 978-1-108-03375-6 Paperback

# STUDIES

— OF —

# PLANT LIFE IN CANADA;

— OR —

## Gleanings from Forest, Lake and Plain,

— BY —

### Mrs C. P. TRAILL,

LAKEFIELD, Ont.,

*Author of "Backwoods of Canada," "Canadian Crusoes," &c., &c.*

Illustrated with Chromo-Lithographs from Drawings by
Mrs. CHAMBERLIN, Ottawa.

**OTTAWA:**
A. S. WOODBURN, PRINTER AND PUBLISHER, ELGIN ST.
1885.

To

His Excellency

The Most Honourable

THE MARQUIS OF LANSDOWNE,

G. C. M. G.,

GOVERNOR GENERAL OF CANADA,

THIS BOOK IS,

By His Excellency's kind Permission,

Respectfully and Gratefully

Dedicated.

# PREFACE.

THIS little work on the FLOWERS and NATIVE PLANTS of Western Canada is offered to the Canadian public with the hope that it may prove a means of awakening a love for the natural productions of the country, and a desire to acquire more knowledge of its resources. It is not a book for the learned. The aim of the writer is simply to show the real pleasure that may be obtained from a habit of observing what is offered to the eye of the traveller,—whether by the wayside path. among the trees of the forest, in the fields, or on the shores of lake and river. Even to know the common name of a flower or fern is something added to our stock of knowledge, and inclines us to wish to know something beyond the mere name. Curiosity is awakened, and from this first step we go on to seek for higher knowledge, which may be found in works of a class far above what the writer of the present book can aspire to offer to the reader. The writer has adopted a familiar style in her descriptions of the plants, thinking it might prove more useful and interesting to the general reader, especially to the young, and thus find a place on the book-shelves of many who would only regard it for the sake of its being a pretty, attractive volume, on account of the illustrations. These, indeed, are contributed by the pencil of a gifted and accomplished lady, Mrs. AGNES CHAMBERLIN, a beloved relative, to whose artistic taste and talents the authoress is greatly indebted. She is conscious that many imperfections will be found in this volume, the contents of which have been written at intervals during a long series of years, many of which were marked by trials, such as fell to the lot of the early colonist and backwoods settlers, and others of a more afflicting nature, which required patience and faith to bear, and to say " Thy Will be done, O Lord."

There is a common little weed that is known by the familiar name of Carpetweed, a small Polygonum, that grows at our doors and often troubles us to root up, from its persevering habits and wiry roots. It is crushed by the foot and bruised, but springs again as if unharmed beneath our tread, and flourishes under all circumstances, however adverse. This little plant had lessons to teach me, and gave courage when trials pressed hard upon me. The simplest weed may thus give strength if we use the lesson rightly, and look up to Him who has pointed us to that love which has so clothed the grass of the field and cared for the preservation of even the lowliest of the herbs and weeds. Will He not also care for the creature made in His own image ? Such are the teachings which Christ gave when on earth. Such teachings are still taught by the flowers of the field.

Mothers of Canada, teach your children to know and love the wild flowers springing in their path, to love the soil in which God's hand has planted them, and in all their after wanderings through the world their hearts will turn back with loving reverence to the land of their birth : to that dear country, endeared to them by the remembrance of the wild flowers which they plucked in the happy days of childhood.

As civilization extends through the Dominion, and the cultivation of the tracts of forest land and prairie, destroys the native trees and the plants that are sheltered by them, many of our beautiful wild flowers, shrubs and ferns will, in the course of time, disappear from the face of the earth, and be forgotten, it seemed a pity that no record of their beauties and uses should be preserved ; and as there is no national botanical garden in Canada, where collections of the most remarkable of our native plants might be cultivated and rescued from oblivion, any addition to the natural history of the country that supplies this want is therefore not without its value to the literature and advancement of the country, and it is hoped that it may prove valuable to the incoming immigrant who makes Canada an abiding home.

Mrs. TRAILL takes this opportunity of acknowledging the kind and invaluable assistance which she has received from her friend, Mr. JAMES FLETCHER, of the Dominion Library, and the encouragement to her labours by Professor MACOUN's opinion of the usefulness of her work on the vegetable productions of the country. She has also to acknowledge the benefit derived from the pamphlet on the "Canadian Forest Trees," by her respected friend, Dr. HURLBURT.

The book is divided, for greater convenience, into four parts :— " The Wild, or Native Flowers " : " Flowering Shrubs " ; " Forest Trees," and " Ferns."

The Wild, or Native Flowers and Flowering Shrubs, are arranged. as a general rule, in the order in which they appear in the woods : but it has been thought that by grouping them somewhat in families, especially where only a short mention is made of some species, it would be easier to refer to them, than if this order were strictly adhered to.

C. P. T.

# CONTENTS

## WILD, OR NATIVE FLOWERS.

## The Flowering Shrubs of Central Canada.

## FOREST TREES.

PAGE

## NATIVE FERNS.

PLATE · I.

BLOOD-ROOT (*Sanguinaria Canadensis*)

# INTRODUCTORY PAGES.

" There's nothing left to chance below ;
The Great Eternal cause
Has made all beauteous order flow
From settled laws. "

EVERY plant, flower, and tree has a simple history of its own,
not without its interest if we would read it aright. It forms
a page in the great volume of Nature which lies open before
us, and without it there would be a blank,—in nature there is no
space left unoccupied.

We watch on some breezy day in summer one of the winged
seeds of the Thistle or Dandelion taking its flight upward and onward,
and we know not where it will alight, and we see not the wisdom of
Him

" Who whirls the blowballs' new fledged pride
In mazy rings on high,
Whose downy pinions once untied
Must onward fly.

Each is commissioned, could we trace
The voyage to each decreed,
To convey to some barren place
A pilgrim seed."—*Agnes Strickland.*

When the writer of the little volume now offered to the Canadian
public first settled in the then unbroken back-woods, on the borders
of the Katchawanook, just where the upper waters of a chain of
lakes narrow into the rapids of the wildly beautiful Otonabee ; the
country at that time was an unbroken wilderness. There was no
opened road for the rudest vehicle on the Douro side of the lakes,
and to gain her new home, the authoress had to cross the river at
Auburn, travel through the newly cut out road in the opposite township, and again cross over to the Otonabee at the head of the rapids
in a birch-bark canoe. There was at that period no other mode of

connection with the northern part of the Douro,—now a branch rail-road from Peterboro' terminates in the flourishing village where once the writer wandered among the forest pines looking for wild flowers and ferns.

As to the roads, one might say, with the Highland traveller,

" Had you but seen these roads before they were made
" You'd have lift up your hands and have blessed General Wade."

The only habitations, beyond our own log cabin at the date of which I write, were one shanty, and the log house of a dear, lamented and valued brother, the enterprising pioneer, the founder of that prosperous village of Lakefield.

It may easily be imagined that there were few objects of interest in the woods at that distant period of time—1832—or as a poor Irish woman sorrowfully remarked, " 'Tis a lonesome place for the likes of us poor women folk ; sure there isn't a hap'orth worth the looking at ; there is no nothing, and it's hard to get the bit and the sup to ate and to drink."

Well, I was better off than poor Biddy Fagan, for I soon found beauties in my forest wanderings in the unknown trees and plants of the forest. These things became a great resource, and every flower and shrub and forest tree awakened an interest in my mind, so that I began to thirst for more intimate knowledge of them. They became like dear friends, soothing and cheering, by their sweet unconscious influence, hours of loneliness, and hours of sorrow and suffering.

Having never made botany a study, and having no one to guide and assist me, it was studying under difficulties, by observation only ; but the eye and the ear are great teachers, and memory is a great storehouse in which are laid up things new and old, which may be drawn out for use in after years. It is a book, the leaves of which can be turned over and read from childhood to old age.

Having experienced the need of some familiar work, giving the information respecting the names and habits and the uses of the native plants, I early conceived the idea of turning the little knowledge, which I gleaned from time to time, to supplying a book which I had felt the want of myself ; but I hesitated to enter the field where all I had gathered had merely been from simply studying the subject without any regular systematic knowledge of botany. The only book that I had access to was an old edition of a " North American Flora," by that good and interesting botanist Frederick Pursh. This work was lent to me by a friend, the only person I knew who had paid any attention to botany as a study, and to whom I was deeply indebted for many hints and for the cheering interest that she always took in my writings, herself possessing the advantages of a highly cultivated mind, educated and trained in the society of persons of scientific and literary notoriety in the Old Country. Mrs. Stewart was a member of the celebrated Edgeworth family. " Pursh's Flora," unfortunately for me, was chiefly written in Latin. This was a drawback in acquiring the information I required ; however, I did manage to make some use of the book, and when I came to a standstill, I had recourse to my husband, and there being a dictionary of the common names, as well as one of the botanical, I contrived to get a familiar knowledge of both. My next teachers were old settlers' wives, and choppers and Indians. These gave me knowledge of another kind, and so by slow steps I gleaned my plant lore—but it was under difficulty. Having no resource in botanical works on our native Flora, save what I could glean from Pursh, I relied entirely upon my own powers of observation, and this did very much to enhance my interest in my adopted country and add to my pleasure as a relief, at times, from the home longings that always arise in the heart of the exile, especially when the sweet opening days of Spring return to the memory of the immigrant Canadian settler, when the hedges put out their green buds, and the Violets scent the air, when pale Primroses and the gay starry Celandine gladden the eye, and the little green lanes and wood-paths are so pleasant to ramble through, among the Daisies and Blue-bells, and Buttercups; and all the gay embroidery of English meads and hedgerows put on their bright array. But for the Canadian forest flowers, and trees and shrubs, and the lovely ferns and mosses, I think I should not have been as contented as I have been, away from dear old England. It was in the hope of leading other lonely hearts to enjoy the same pleasant recreation, that I have so often pointed out the natural beauties of this country to their attention, and now present my forest gleanings to them in a simple form, trusting that it may not prove an unacceptable addition to the literature of Canada, and that it may become a household book, as Gilbert White's Natural History of Selborne is to this day among English readers ; and now at the age of eighty-three years, fifty-two of which have been spent in the fair province of Ontario, in her far forest home on the banks of the rapid Otonabee, the writer lays down her pen, with earnest prayers for the prosperity of this her much beloved adopted country; that with the favour and blessing of our God it may become the glory of all lands.

LAKEFIELD, Ont., 1884.

# WILD, OR NATIVE FLOWERS.

---

VIOLETS.

The violet in her greenwood bower
  Where birchen boughs with hazel mingle,
May boast herself the fairest flower
  In forest, glade or copsewood dingle.—*Scott.*

THERE is music and poetry in the very name—" Violet."   In the forest wilderness, far removed from all our early home associations, the word will call up, unbidden, a host of sweet memories of the old familiar land where, as children, we were wont to roam among bowery lanes, and to tread the well-worn pathway through green pastures down by the hawthorn hedge, and grassy banks, where grew in early spring, Primroses, Blue-bells, and purple Violets.   What dainty, sweet smelling posies have you and I, dear reader, (I speak to the emigrants from the dear Old Country) gathered on sunny March and April days on those green banks and grassy meadows?   How many a root full of freshly opened Violets or Primroses, have· we joyfully carried off to plant in our own little bits of garden ground, there to fade and wither beneath the glare of sunshine and drying winds ; but little we heeded, the loss was soon replaced.

And still I doubt not but that Violets and Primroses, the Blue-bells and the Cowslips yet bloom and flourish in the loved haunts of our childhood.   Year after year sees them bloom afresh—pure, sweet and fragrant as when last we filled our laps and bosoms with their flowers or twined them in garlands for our hair : but we change and grow old ; God wills it so, and it is well !   Though Canada boasts of many members of this charming family, there are none among our Violets so deeply blue, or so deliciously fragrant, as the common English March Violet, *Viola odorata.*   This sweet flower bears away the crown from all its fellows.   One of our older poets (Sir Henry Wotton) has said, as if in scorn of it, when compared with the rose,

" Ye violets that first appear,
    " By your pure purple mantles known,
" Like the proud virgins of the year
    " As if the spring were all your own.
    " What are ye—when the rose is blown ? "

Good Sir Henry, we would match the perfume of the lowly violet
even against the fragrance of the blushing rose.

Though deficient in the scent of the purple Violet of Europe, we
have many lovely species among the native Violets of Canada.   The
earliest is the small flowered

### EARLY WHITE VIOLET—*Viola blanda* (Willd).

This blossoms early in April, soon after the disappearance of the
snow.   The light green smooth leaves may be seen breaking through
the black, damp, fibrous mould closely rolled inward at the margins ;
the flowers are small, rather sweet scented, greenish white, with delicate
pencillings of purple at the base of the petals ; it is a moisture-loving
plant, and affects open, recently overflowed ground, near creeks.   It
comes so early that we welcome its appearance thankfully for it

" Tells us that winter, cold winter is past,
And that spring, welcome spring, is returning at last."

On pulling up a thrifty plant, late in the summer, it surprises you
with a new set of flowers, quite different from the spring blossoms ; these
are small, buds and flowers of a dull chocolate-brown, lying almost
covered over in the mould, with seed pods, some ready to shed the
ripened seed, others just formed.   This mysterious little plant has
been distinguished by some botanists as *Viola clandestina*, from the
curious hidden way in which it produces the subterranean flowers and
seeds ; others have considered it as identical with the next species.

### THE PENCILLED VIOLET, *Viola renifolia*, (Gray),

which bears its white blossoms on rather long slender foot-stalks, and
which are slightly larger than those of the above, milky white with dark
veinings.   The leaves, although covered with soft hairs, have a curious
smooth and shining appearance.   They are round heart or kidney-shaped,
notched at the edges.   As the summer advances the foliage of the
Pencilled Violet increases in luxuriance and many white fibrous running
roots are produced in the loose soil.   This attractive species may be
found in swamps and forests, growing amidst decayed wood and mosses,
and increasing after the same manner as *Viola blanda*.   A point which
easily distinguishes this species from the last is the total absence of
scent ; the leaves, too, are much more pubescent—a character which is
very noticeable in the early morning when they are covered with dew.

The commonest among our blue Violets is

THE HOODED VIOLET—*Viola cucullata,* (Ait.)

so called from the involute habit of the leaves, which, when first appearing, are folded inwardly as if to shield the tender buds of the flowers from the chilling winds. There are many forms or varieties of this species varying very much in appearance, the difference being probably due to the habitat in which they occur. One of the hand-somest is the Large Blue Wood Violet, which flowers about the middle of June, has blue scentless flowers with round petals, and large blunt hirsute leaves, and is found in low woods. Another variety, with deep violet flowers, has elongated petals and pointed, rather smooth, leaves of a purplish tint, at least till late in the season. It is found on open sunny banks, and dry grassy hill-sides. Yet another variety is often found by the sides of springs and rivers, forming spreading tufts among the grass with its smooth pointed leaves and pale delicate flowers.

The prettiest of all our blue Violets is the

ARROW-LEAVED VIOLET—*Viola sagittata,* (Ait.)

It is found in low, sandy, shady valleys or very light loamy soil. The leaves of this species are not always arrow or heart-shaped, but also long and narrow, blunt at the apex, decurrent on the short leaf-stalk, notched at the edges, and rather roughened and dulled in colour by the short silvery hairs on the surface. The flowers rise singly from the crown of the plant; colour—a full azure-blue, a little white at the base of the petals which are bearded with soft silky wool; anthers—a bright orange, which form a tiny cone from the meeting of the tips. The flowers, six or eight in number, fall back from the centre, and lie prostrate on the closely horizontal leaves. The unopened buds are sharply folded with bright green sepals, and are of a deep bluish-purple. Another form, sometimes called *Viola ovata,* very nearly resembles the above, but the leaves are less hairy, and the color is more purple in the tint.

Among the white Violets none are so beautiful as the

BRANCHING WHITE WOOD VIOLET—*Viola Canadensis,* (L.)

This, our Canada Violet, is worthy of a place in the garden. Not only is it a lovely flower, but it takes kindly to garden culture, preferring a shady place to the open sunshine. In its native haunts, the rich black vegetable mould of beech and maple woods, it rises to the height of from nine inches to a foot, throwing out slender leafy-bracted branches, with many buds and pure milk-white flowers. The petals are slightly clouded on the outside with purple : the buds are also dark, while the petals of the flower are veined with purple, and in some cases there is a shade of yellow in the centre of the flowers, but this is not seen when under cultivation.

The plant continues to send forth blossoms all through the summer and even late in the month of September when undisturbed, the seeds ripening early, form new plants, which, sheltered by the parent stems, continue to increase, forming a compact ball of snow-white flowers. This has been the case in my own garden. If well watered and in suitable soil, this pretty branching violet may be taken from the woods even in full bloom, and will grow and continue to blossom freely, but must have shade and moisture and leaf-mould to ensure success to its healthy growth. The leaves are large, broad at the base, narrowing to a very slender point and coarsely toothed.

Among the branching Violets we have two very pretty lilac ones : Long Spurred Violet (*Viola rostrata*) and the Dog Violet (*Viola canina var. sylvestris*). These pretty species are distinguished by the long spur, lilac-tinted petals, striped and veined with dark purple and branching stem. The next in point of interest is the

DOWNY YELLOW VIOLET—*Viola pubescens*, (Ait.)

This handsome species is confined to our forests and copses. It will attain to more than a foot in height in its rich native woods, and blossoms in early summer ; the colour is golden yellow, veined with black jetty lines. The seed-vessels are deeply clothed with white silky wool.

The Yellow Violet has been immortalised by the sweet verses of that rare poet of nature Cullen Byrant—almost every child is familiar with his stanzas to the Yellow Violet. There is another variety of this Violet, called *var. scabriuscula*, which is not so branching, lower, the leaves darker, and and the blossoms smaller but of a deeper golden colour. This variety is found in drier, more open soil—the black veining more distinctly marked than in the downy Yellow Violet, and the seed-vessels smooth. They both improve under culture, having two sets of flowers during the season.

The Violet has ever been a favourite flower with the poets, from Shakespeare and Milton down to the present day we find mention of this lovely flower scattered through their verses. Nor are the old Italian poets silent in its praise. Luigi de Gonzaga, in stanzas addressed to his lady-love, says :—

> " But only violets shall twine
> Thy ebon tresses, lady mine."

Milton in his sonnet to Echo speaks of the " Violet embroidered vale."

Here are lines to early Violets after the manner of the old English poet Herrick :—

> Children of sweetest birth,
> Why do ye bend to earth
> Eyes in whose deepest blue
> Sees but the diamond dew.
>
> Has not the early ray
> Yet kissed those tears away
> That fell with closing day ?
> Say do ye fear to meet
> The hail and driving sleet
> Which gloomy winter stern,
> Flings from his snow-wreathed urn ?
>
> Or do ye fear the breeze,
> So sadly sighing thro' the trees.
> Will chill your fragrant flowers,
> 'Ere April's silv'ry showers
> Have visited your bowers ?
> Why came ye till the Cuckoo's voice
> Bade hill and dale rejoice ?
> Till Philomel with tender tone
> Waking the echoes lone,
> Bade woodland glades prolong
> Her sweetly tuneful song.
>
> Till Sky-lark blithe, and Linnet grey,
> From fallow brown and meadow gay,
> Pour forth their jocund roundelay.
> Till Cowslips wan, and Dasies pied,
> Broider the hillocks side :
> And opening Hawthorn buds are seen
> Decking the hedge-row screen.
>
> What though the Primrose drest
> In her pure modest vest,
> Come rashly forth
> To brave the biting North.
> Did ye not see her fall
> Straight 'neath his snowy pall ?
> And heard ye not the West wind sigh
> Her requiem as he hurried by ?
>
> Go hide ye then till groves are green
> And April's clouded bow is seen,
> Till suns are bright, and skies are clear,
> And every flower that doth appear
> Proclaims the birthday of the year.—*C.P. T., Lakefield*

### Liver-Leaf—Wind-Flower—*Hepatica acutiloba,* (D.C.)

" Lodged in sunny clefts,
    Where the cold breeze comes not, blooms alone
    The little Wind-flower, whose just opened eye[*]
    Is blue, as the spring heaven it gazes at."—*Bryant.*

The American poet, Bryant, has many happy allusions to the Hepatica under the name of " Wind-flower." The more common name among our Canadian settlers, is " Snow-flower," it being the first blossom that appears directly after the melting off of the winter snows.

In the forest—in open, grassy woods, on banks and upturned roots of trees—this sweet flower gladdens the eye with its cheerful, starry blossoms ; every child knows it, and fills its hands and bosom with its flowers—pink, blue, deep azure and pure white. What the Daisy is to England, the Snow-flower or Liver-leaf is to Canada. It lingers long within the forest shade, coyly retreating within its sheltering glades from the open glare of the sun : though for a time it will not refuse to bloom within the garden borders, when transplanted early in spring, and doubtless if properly supplied with black mould from the woods and partially sheltered by shrubs it would continue to grow and flourish with us constantly.

We have two sorts, *H. acutiloba* and *H. triloba.* A large variety has been found on Long Island in Rice Lake, the leaves of which are five lobed; the lobes are much rounded, the leaf stalks stout, densely silky, the flowers large, of a deep purple blue. This handsome plant throve under careful cultivation, and proved highly ornamental.

The small, round, closely-folded buds of the Hepatica, appear before the white, silky, leaves unfold themselves, though many of the old leaves of the former year remain persistent through the winter. The buds rise from the centre of a silken bed of soft sheaths and young leaves, as if nature kindly provided for the warmth and protection of these early flowers with parental care.

Later in the season, the young leaves expand, just before the flowers drop off. The white flowered is the most common among our Hepaticas, but varieties may be seen of many hues—waxen pink, pale blue and azure blue with intermediate shades and tints.

The Hepatica belongs to the Nat. Ord. Ranunculaceæ, the Crowfoot family, but possesses none of the acrid and poisonous qualities of the Ranunculus proper, being used in medicine, as a mild tonic, by the American herb doctors in fevers and disorders of the liver.

It is very probable that its healing virtues in complaints of the liver, gave rise to its common name in old times ; some assign the name, " Liver-leaf," to the form of the lobed leaf.

---

[*] The-blue flowered "Hepatica triloba" is evidently the flower meant by the poet.

BLOOD-ROOT.—*Sanguinaria Canadensis*, (L.)

(PLATE I.)

" Here the quick-footed wolf
Pausing to lap thy waters, crushed the flower
Of Sanguinaria, from whose brittle stem
The red drops fell like blood."—*Bryant.*

Just at the margin of the forest, and in newly cleared ground among the rich black leaf mould, may be seen late in April and May the closely folded vine-shaped leaf of the Blood-root, enclosing in its fold one pure white bud.

The leaf is strongly veined beneath with pale orange veins, the simple semi-transparent round leaf stalk as well as the flower scape, is filled with a liquor of a bright orange red colour : break the thick fleshy tuberous root and a red fluid drops from every wounded pore, whence its local name " Blood-root."

This juice is used largely by the Indian squaws in their various manufactures. With it they dye the porcupine quills and moose-hair both red and orange, and also stain the baskets of a better sort that they offer for sale in the stores. Nor is this the only use to which it is applied : they use the juice both externally in curing cutaneous eruptions of the skin, and internally in other diseases. Latterly its medicinal qualities have been acknowledged by the American Eclectic School of Pharmacy as valuable in many forms of disease, so that we find our beautiful plant to be both useful and ornamental.

The Blood-root grows in large beds ; each knob of the root sends up one leaf, and its accompanying flower bud which it kindly enfolds as if to protect the fair frail blossom from the chilling winds and showers of hail and sleet. The leaf is of a greyish or blueish green, at first the underside, which is the part exposed to view, is salmon coloured veined with red, but as it expands and enlarges the outer surface darkens into deeper green. The blossom is composed of many petals, varying from eight to twelve. The many stamens are of a bright orange yellow. The stigma is two-lobed, and the style short or sessile. The seed is contained in an oblong pod of two valves. The seeds are of a bright red brown colour. The ivory white petals are oblong, blunt, or sometimes pointed ; the inner ones narrower than the outer, at first concave, but opening out as the flower matures. Under cultivation the blossom of the Blood-root increases in size, but the plant does not seem to spread and multiply freely as in its native soil. It is one of our most lovely native Spring flowers. It is a pity that, with the march of civilization, we shall soon lose its fair pure blossoms. It is easily cultivated, and repays care by the increase in size of the flowers, ripening the seeds perfectly and freely.

### TALL BUTTERCUP—*Ranunculus acris*, (L.)

We see the old familiar meadow-flower of our childhood bright and gay, growing abundantly in low, wet pasture lands in Canada, where it becomes to the eye of the farmer a troublesome, unprofitable weed, rejected by the cattle for its bitter acrid qualities. Yet it is pleasant to meet its old familiar face in a foreign land, where often the sight of some simple flower will awaken tender recollections of early scenes of sunny grassy meadows, where we wandered in days of thoughtless childhood, free of care as the Lark that carolled above our heads in the glad sunshine; happy days brought back, in all their freshness, to memory by the sight of a simple yellow Buttercup blossoming in Canadian wilds and wastes : despised and rejected by others, but precious to the heart of the lonely immigrant who hails it as a tiny link between himself and his early home life.

### EARLY CROWFOOT—*Ranunculus fascicularis*, (Muhl.)

This native species of Ranunculus is one of our earliest spring flowers. It grows low and spreading to the ground, the foliage, hairy, which gives a hoary tint to the divided coarsely cut leaves ; the blossoms are of a pale yellow colour, not as large as the common Buttercup. The root is a cluster of thick, fleshy fibres.

One of the prettiest of the Ranunculus family is the

### CREEPING SPEAR-WORT—*Ranunculus reptans*, (Gray.)

a tiny delicate plant, with slender thready stems rooting from beneath the joints. The leaves are very narrow, and pointed, those nearest to the root a little lobed or eared. The little bright, golden, shining flowers, only a few lines broad, are borne in the axils of the leaves of the prostrate creeping stems, and peep out from the sandy soil among tufts of minute hairy sedges (*Eleocharis acicularis*) that clothe the damp low-lying shores of rivers or lakes. There are several Water Crow-foots, some with white flowers, others with yellow. These flowers float upon the surface of still-flowing rivers or lakes, gently rising or falling with the motion of the waters. The beautiful adaptation of plants to soil and circumstances may be noticed in these and some other aquatic plants which have their foliage dissected into narrow segments, so that the water may freely flow through them. Of the water Ranunculi, we may mention White Water Crowfoot (*R. aquatilis*) and Yellow Water Crowfoot (*R. multifidus*).

There are among our native Ranunculus flowers a few plants of which the outward beauties of their blossoms are better known to us

than their useful qualities, though doubtless even the lowliest among
them has a part to perform, though not apparently for man's sole benefit
but also for the support or shelter of some of God's creation among the
insect tribes or smaller animals or birds which find nourishment in their
seeds, leaves or roots. It is a remarkable fact but it is rarely, if ever
the case, that the flower is selected of any plant for food by bird
or beast.

There are many native plants of the order Ranunculaceæ,
too many to be here described. Gray describes nineteen species of
Ranunculus proper, only a part of the plants described being found with
us, and there are doubtless many others found in our extensive
Dominion not at present named. The large, deep golden, abundant
flowers of the

### MARSH MARIGOLD—*Caltha palustris*, (L.)

are too well known to need any minute description. It is, indeed, a
splendid flower, and can hardly fail of being admired, when seen, like a
" field of cloth of gold," covering the low, wet ground with its large leaves
of a deep refreshing green, and its rich golden cups : a pleasant sight to
the eye in May. The leaves were used as a pot-herb by the early
backwoods settlers, before gardens were planted, but through carelessness
or ignorance, accidents of a fatal nature are known to have occurred
by gathering the leaves of the *Arisæma triphyllum* with those of the
more innocent herb the Marsh Marigold, or Water Cowslip, as this
plant is often called.

### MITREWORT, BISHOP'S CAP.—*Mitella diphylla*, (L.)

This elegant forest flower is found in moist rich soil among beech,
maple, and other hardwood trees.

We have two species of these plants : one *Mitella nuda*, *L.*, rather
creeping, with green blossoms, only a few inches in height, and the
flowers larger and fewer on the slender scape, the bright green lobed
leaves spreading on the ground. The taller Mitrewort has elegant
fringed cups, greenish white, many flowers arranged in a long slender
spike. The term diphylla distinguishes it from the low dwarf species,
there being two opposite pointed leafy bracts about the middle of the
long slender scape. Not only are the fringed cap-like flowers worthy of
minute attention, but the boat-shaped two-valved capsules of the seed
vessels form a pretty feature in the plant. At an early stage of ripeness
the shining jet black seeds appear and are scarcely less attractive than

the delicate fringed flowers, and have given rise to the local name in some places of "Gem-flower." Nearly allied to the above is the woodland flower

### FALSE MITREWORT.—*Tiarella cordifolia*, (L.)

to which the name Wood Mignonette is often given, not with respect to its scent, for there is no particularly agreeable odour in the flower, and the leaves are somewhat coarse and pungent in quality ; but for the beauty of the light graceful blossoms which are white with orange tipped or light tawny brown anthers. The petals are pointed and five in number ; stamens ten, long and slender ; styles, two ; seed vessel, two valved ; the base of the pistil is thickened, forming a turban-like pod.

There are two forms of our pretty "Wood Mignonette"—one with closer, more globular, heads of flowers, the other with the flowers looser and more scattered. Both affect the rich black mould and shade of the forest trees.

The plant might be called evergreen, as the leaves appear green and fresh from beneath the covering of Winter's snow. The large flat sharply-toothed, lobed, leaves are shaded in the centre with purple ; the veinings also blackish purple, and the surface is beset with very short appressed hairs. The leaf stalks of the young plants are of a reddish pink and hairy at their junction with the root.

### WOOD BETONY—*Pedicularis Canadensis*, (L.)

This plant is commonly found in open grassy thickets and plainlands. Of the two common species, we have one with dark, dull red flowers, and another with yellow. It is a rather coarse flower ; the spike leafy, hairy and rough : the leaves are divided into many rounded lobes, toothed at the margins, and deeply cleft, nearly to the mid-rib, turning black in drying. The yellow flowered is a smaller plant than the red ; the foliage is much more hairy, and the lipped blossoms are also hairy, the upper lip arched over the lower lobes of the corolla. I think it must be a distinct variety, or even species. Lindley remarks in his "Natural System,' that the Betony is acrid in quality but that it is eaten by goats : unluckily we have no goats in Canada to benefit by the herbage of this homely plant.

### FLOWERING WINTER-GREEN—*Polygala paucifolia*, (Willd.)

This is one of our early flowering plants, distinguished by the common name of "Winter-green." It belongs to a family of well-known plants called Milk-worts—low, bitter herbs—some of which are remarkable for tonic properties, of which the Senega, or Snake-root, is an example.

Some of the species are remarked as bearing fertile flowers under ground. The flowers of some are white, some red and others purple or reddish lilac. The name Milk-wort appears to have been adopted without any foundation, from an imaginary idea that the herbage of some of the species promoted the secretion of milk in cows. Several of the milk-worts are indigenous to Canada.

*P. Senega* is not evergreen in its habit ; it flowers in May among grasses on dry uplands ; it is simple, slender, and not ungraceful, the leafy stem terminating in a spike of greenish-white flowers. The wiry root is said to possess medicinal qualities. The plant which merits our attention more particularly for its beautiful flowers is *P. paucifolia*, the beautiful fringed, or crested, Polygala. It is a small-sized plant, about six to nine inches in height ; the stem is simple, rising from a running or creeping root-stalk, often furnished with subterranean imperfect leaflets, and fertile flowers. The smooth, dark-green leaves, delicately fringed with soft, silky hairs, tinged with a purplish hue, are persistent through the winter. The stem of the plant is leafy : the lower leaves small and bract-like, the upper ones larger and clustered round the summit ; from amongst these appear from two to four, and sometimes as many as five elegantly winged purple-lilac flowers. The two upper petals are long-ovate, the lower forming a crested keel, finely tinged with deeper purple. The flowers of this beautiful species are very graceful, slightly drooping from among the shining leaves on thread-like pedicels. The stamens are six ; sepals of the calyx, five ; petals, three. Some old writers have given the name of " Fly-flower " to our pretty Polygala, and truly not an inappropriate name, as one might not inaptly liken the opened blossom to some gay purple-winged insect ready to take its flight from the bosom of the soft silky leaves that form an involucre round it.

This Flowering Wintergreen is one of our earliest spring flowers ; in fine warm seasons it appears in the latter end of April, continuing to bloom on till the middle of May. The early flowering plants are not so tall, neither are the flowers so large as those put forth later in the season. On sunny spots, on moderately sandy soil, on open waste, by the wayside, or at the edge of the partly cleared forest, it expands its soft purple— sometimes rose-coloured flowers—often mingled promiscuously with the white blossoms of the wild Strawberry and creeping Early Everlasting. The lovely winged flowers gladden the eye of the traveller, when as yet but few blossoms have ventured to brave the late frosts that oftimes nip the fair promise of the Spring.

No wonder that we watch with pleasure for the re-appearing of our little floral gem, as in old times we did for the bright golden varnished flowers of the Smaller Celandine, that starred the green turfy banks in our English lanes, opening so gaily to the ruffling winds and sunshine on bright March mornings. Some of the peasants and old writers call the little Celandines—"Kingcups"—and I have often fancied that Shakespeare was thinking of this sweet spring flower when he wrote his charming song, Hark the lark at Heaven's gate sings,

> " And winking *Mary-buds* begin
> To ope their golden eyes ;
> With all the things that pretty bin,
> My lady sweet arise. "

Mary-golds, which some suppose the poet meant by *Mary-buds*, have little poetical charm about them, not being associated with the Lark, as a wild spring-flower. It is more than probable it was the gay little Celandine that he thus immortalises with his sweet song.

The larger form of our Flowering Wintergreen is found somewhat later in May, in the woods, and is known by the settlers as "Satin-flower." It would make a pretty border plant, and from its early flowering would be a great acquisition to our gardens.

AMERICAN SNAKE-ROOT.—*Polygala Senega, (L.*

already referred to, is less ornamental, though a delicate and graceful little plant. Like the rest of the genus its root is perennial, woody and bitter in its qualities. The stem is simple, wand-like, clothed with lanceolate leaves, and terminating in a spike of greenish white flowers. The wings of this species are small, and embrace the flattened less conspicuously crested keel. Its favourite haunt is dry upland plains, among shrubs and wild grasses ; it blossoms later than the more showy purple Polygala, being seen through May and June.

Another purple-flowered species is

SLENDER PURPLE MILKWORT.—*Polygala polygama*, (Walt.)

The flowers form slender racemes of violet coloured flowers springing from a woody root-stock, which also bears numerous inconspicuous, but more fertile flowers, beneath the ground. Its usual habitat is dry grassy banks, in sandy or rocky ravines ; all these plants seem to prefer sunshine to shade, and a light sandy, loamy soil. Several of the species are used as tonics and alteratives by the American herbalists

WOOD ANEMONE—*Anemone nemorosa*, (Lin.)

" Within the wood,
Whose young and half transparent leaves,
Scarce cast a shade ; gay circles of Anemones,
Danced on their stalks."—*Bryant.*

The classical name Anemone is derived from a Greek word, which signifies the wind, because it was thought that the flower opened out its blossoms only when the wind was blowing. Whatever the habits of the Anemone of the Grecian Isles may be, assuredly in their native haunts in this country, the blossoms open alike in windy weather or in calm ; in shade or in sunshine. It is more likely that the wind acting upon the downy seeds of some species and dispersing them abroad, has been the origin of the idea, and has given birth to the popular name which poets have made familiar to the ear with many sweet lines. Byrant who is the American poet of Nature, for he seems to revel in all that is fair among the flowers and streams and rocks and forest shades, has also given the name of " Wind-flower" to the blue Hepatica.

This pretty delicate species loves the moderate shade of groves and thickets ; it is often found in open pinelands of second growth, and evidently prefers a light and somewhat sandy soil to any other ; with glimpses of sunshine stealing down upon it.

The Wood Anemone is from four to nine inches in height, but occasionally taller ; the five rounded sepals which form the flower are white, tinged with a purplish-red or dull pink on the outside. The leaves are three—parted, divided again into three, toothed and sharply cut, and somewhat coarse in texture ; the three upper stem leaves form an involucre about midway between the root and the flower-cup.

Our Wood Anemone is a cheerful little flower, gladdening us with its blossoms early in the month of May. It is very abundant in the neighbourhood of Toronto, on the grassy banks and piny-dells at Dover Court, and elsewhere.

" There thickly strewn in woodland bowers,
Anemones their stars unfold."

A taller species, *Anemone dichotoma*, with very beautiful white starry flowers, is found on gravelly banks by river-sides and under the shade of shrubs in most parts of Canada, as is, also, the downy seeded species known as " Thimble-weed" *Anemone cylindrica* from the cylindrical heads of fruit. This latter is not very attractive for beauty of colour ; the flower is greenish-white, small, two of the sepals being shorter and less conspicuous than the others. The plant is from one to two feet high ; the leaves of the cut and pointed involucre are

B

coarse; of a dull green, surrounding the several long flower-stalks. The soft cottony seeds remain in close heads through the winter, till the spring breezes disperse them.

The largest species of our native Anemones is the Tall Anemone, *A. Virginiana.* This handsome plant loves the shores of lakes and streams; damp rich ground suits it well, as it grows freely in such soil, and under moderate shade when transferred to the garden.

The foliage of the Tall Anemone is coarse, growing in whorls round the stem; divisions of the leaf, three parted, sharply pointed and toothed. In this, as in all the species, the coloured sepals, (or calyx leaves) form the flower. The outer surface of the ivory-white flower is covered with minute silky hairs, the round flattened silky buds rise singly on tall naked stems; but those of th outer series are supplied with two small leaflets embracing the stalk   The central and largest flowers open first the lateral or outer ones, as these fade away; thus a succession of blossoms is produced, which continue to bloom for several weeks.   The flowers of this plant, under cultivation, become larger and handsomer than in their wild state.   This species is distinguished from *A. cylindrica* by its round heads of fruit and larger flowers.   The Anemone is always a favourite flower wherever it may be seen, whether in British woods, on Alpine heights, or in Canadian wilds; on banks of lonely lakes and forest streams, or in the garden parterre, where it is rivalled by few other flowers in grace of form or splendour of colour.

We cannot boast, in this part of the Dominion, any of the more brilliant and beautiful flowers of this ornamental family, though that interesting lovely species, known as Pasque-flower, *Anemone patens*, (L.) *var Nuttalliana* (Gray), is largely distributed over the prairie lands of the Western States and in our North-Western Provinces, where it is one of the earliest of the Spring flowers to gladden the earth, with its large azure-blue blossoms, than which none are more beautiful.

The bud appears on a thick leafless scape, about four to six inches high, enclosed in a cut and pointed involucre of grey bracts of silvery hue and shining brightness.   The scape is clothed with hairy scales; from within this silky covering peeps out the fair blue bud, which shortly expands into a large, open, bell-like, very blue blossom, with a shade of white at the base of each large pointed sepal.   As the flower advances a change takes place in the whole aspect of the plant: the root-leaves begin to appear, which are compoundly cut and divided, and the head of plumy fruit is raised on a high scape above the silken involucre, and now ripens in the breezy air and sunshine.

I have a fine, perfect, dried specimen before me, under all its several aspects, and wish that it could be oftener seen as a cultivated border ornament in our Canadian gardens. The name "Pasque-flower' is hardly known among the inhabitants of our North-Western prairies, and the Indian name I have not yet obtained ; it would, I am sure, be descriptive of some natural quality of the plant—its growth or habits.

We have in Ontario several distinct species of Anemone, though none so finely coloured as the Prairie flower : nor can we boast of the splendid Anemones that gem the wilderness tracts of Palestine. Some travellers have suggested that it was to the brilliant blossoms of the scarlet, blue, and white Anemones that the Saviour drew the attention of his disciples, while Sir James Smith has supposed—and with more probability—it was to the glowing colours of the golden flowered *Amaryllis lutea*, which abounds on the fields of Palestine, that He alluded in His words—" Behold the Lilies of the field," etc.

SPRING BEAUTY—*Claytonia Virginica* (Lin.) and *C. Caroliniana* (Michx )

> Where the fire had smoked and smouldered,
> Saw the earliest flower of Spring-time,
> Saw the Beauty of the Spring-time,
> Saw the Miskodeed (*) in blossom.—*Hiawatha*.

This simple, delicate little plant is one of our earliest April flowers. In warm springs it is almost exclusively an April flower, but in cold and backward seasons, it often delays its blossoming time till May.

Partially hidden beneath the shelter of old decaying timbers and fallen boughs, its pretty pink buds peep shyly forth. It is often found in partially cleared beech-woods, and in rich moist meadows.

In Canada, there are two species ; *C. Caroliniana*, with few flowers, white, veined with red, and both leaves and flowers larger than the more common western form, *C. Virginica*, the blossoms of which are more numerous, smaller, and pink, veined with lines of a deeper rose colour, forming a slender raceme ; sometimes the little pedicels or flower stalks are bent or twisted to one side, so as to throw the flowers all in one direction.

The scape springs from a small deep tuber, bearing a single pair of soft, oily, succulent leaves. In the white flowered species, *C. Caroliniana*, these leaves are placed about midway up the stem, but in the pink *( C. Virginica )* the leaves lie closer to the ground, and are smaller and narrower, of a dark bluish green hue. Our Spring Beauties well deserve their pretty poetical name. They come in with the Robin and the Song Sparrow, the Hepatica, and the first white Violet ; they linger in shady spots, as if unwilling to desert us till more sunny days

---

(*) Miskodeed—Indian name for Spring Beauty.

have wakened up a wealth of brighter blossoms to gladden the eye ; yet the first, and the last, are apt to be most prized by us, with flowers as well as other treasures.

How infinitely wise and merciful are the arrangements of the Great Creator. Let us instance the connection between Bees and Flowers. In cold climates the former lie torpid, or nearly so during the long months of winter, until the genial rays of the sun and light have quickened vegetation into activity, and buds and blossoms open, containing the nutriment necessary for this busy insect tribe.

The Bees seem made for the Blossoms ; the Blossoms for the Bees.

On a bright March morning what sound can be more in harmony with the sunshine and blue skies, than the murmuring of the honey-bees, in a border of cloth of gold Crocuses ? What sight more cheerful to the eye ? But I forget. Canada has few of these sunny flowers, and no March days like those that woo the hive bees from their winter dormitories. And even April is with us only a name. We have no April, month of rainbows, suns, and showers. We miss the deep blue skies, and silver throne-like clouds that cast their fleeting shadows over the tender springing grass and corn ; we have no mossy lanes odorous with blue Violets.

But our April flowers are comparatively speaking, few, and so we prize our early Violets, Hepaticas and Spring Beauties.

We miss the turfy banks, studded with starry Daisies, pale Primroses and azure Blue-bells.

In the warmth and shelter of the forest, vegetation appears. The black leaf-mould, so light and rich, quickens the seedlings into rapid growth, and green leaves and opening buds follow soon after the melting of the snows of winter. The starry blossoms of the Spring plants come forth and are followed by many a lovely flower, increasing with the more genial seasons of May and June.

Our May is bright and sunny, more like to the English March ; it is indeed a month of promise—a month of many flowers. But too often its fair buds and blossoms are nipped by frost, " and winter, lingering, chills the lap of May."

INDIAN TURNIP.—*Arisæma triphyllum.* (Torr.)

"Or peers the Arum from its spotted veil."—*Bryant.*

There are two species of Arum found in Canada, the larger of which is known as Green-Dragon (*A. Dracontium*) ; the other is known by the familiar name of Indian Turnip (*A. triphyllum* or *A purpureum*).

These moisture-loving plants are chiefly to be found in rich, black, swampy mould, beneath the shade of trees and rank herbage, near creeks and damp places, in or about the forest.

The sheath that envelopes and protects the spadix, or central column which supports the clustered flowers and fruit, is an incurved membranaceous hood, of a pale green colour, beautifully striped with dark purple or brownish-purple. The flowers are inconspicuous, hidden at the base of the scape by the sheath. They are of two kinds, the sterile and fertile, the former, placed above the latter, consisting of whorls of four or more stamens, and two to four-celled anthers, the fertile or fruit-bearing flowers, of one-celled ovaries. The fruit, when ripe, is bright scarlet, clustered round the lower part of the round, fleshy, scape. As the berries ripen, the hood, or sheath, withers and shrivels away to admit the ripening rays of heat and light to the fruit.

The root of the Indian Turnip consists of a round, wrinkled, fleshy corm, sometimes over two inches in diameter ; from this rises the simple scape or stem of the plant, which is sheathed by the base of the leaves. These are on long naked stalks, divided into three ovate pointed leaflets, waved at the edges.

The juices of the Indian Turnip are hot, acrid, and of a poisonous quality, but can be rendered useful and harmless by the action of heat ; the roots roasted in the fire are no longer poisonous. The Indian herbalists use the Indian Turnip in medicine as a remedy in violent colic, long experience having taught them in what manner to employ this dangerous root.

The Arisæma belongs to the natural order *Araceæ*, most plants of which contain an acrid poison, yet under proper care they can be made valuable articles of food. Among these we may mention the roots of *Colocosia mucronatum*, and others, which, under the more familiar names of Eddoes and Yams, are in common use in tropical countries. (Lindley.)

The juice of *A. triphyllum*, our Indian Turnip, has been used boiled in milk, as a remedy for consumption.

Portland Sago is prepared from a larger species, *Arum maculatum*, Spotted Arum The corm, or root, yields a fine, white, starchy powder, similar to Arrow-root, and is prepared much in the same way as Potato starch. The pulp, after being ground or pounded, is thrown into clean water and stirred ; after settling, the water is poured off, and the white sediment is again submitted to the same process until it becomes quite pure and is then dried. A pound of this starch may be made from a peck of the roots. The roots should be dried in sand before using. Thus purified and divested of its poisonous qualities, the powder so procured becomes a pleasant and valuable article of food, and is sold under the name of Portland Sago, or Portland Arrow-root.

When deprived of the poisonous acrid juices that pervade them, all our known species may be rendered valuable both as food and medicine; but they should not be employed without care and experience.

There seems in the vegetable world, as well as in the moral, two opposite principles, the good and the evil. The gracious God has given to man the power, by the cultivation of his intellect, to elict the good and useful, separating it from the vile and injurious, thus turning that into a blessing which would otherwise be a curse.

"The Arum family possesses many valuable medicinal qualities," says Dr. Charles Lee, " but would nevertheless become dangerous poisons in the hands of ignorant persons."

The useful Cassava, *Jatropha manihot* (Lin.), of the West Indies and tropical America, is another remarkable instance of Art overcoming Nature, and obtaining a positive good from that which in its natural state is evil. The Cassava, from the flour of which the bread made by the natives is manufactured, being the starchy parts of a ·poisonous plant of the Euphorbia family, the milky juice of which is highly acrid and poisonous. The pleasant and useful article sold in the shops under the name of Tapioca is also made from the Cassava root.

How well do I recall to mind the old English Arum, known by its familiar names among the Suffolk peasantry as " Cuckoo-pint," " Jack in the Pulpit" and "Lords and Ladies." The first name no doubt was suggested from the appearance of the plant about the time of the coming of that herald of spring the Cuckoo ; the hooded spathe shrouding the spadix like a monkish cowl the second ; while the distinction in the colour between the deep purplish-red and creamy white of the central column or spadix, supplied the more euphonious term of " Lords and Ladies," which to our childish fancies represented the masculine and feminine element in the plant ; of course we dreamed not of the Linnæan system ; the one was the Lord because it was dark, the other the Lady because it was fair and more delicate. This was plain reasoning of the cause ; children never reason, they only see effects. I am afraid that in many things I am yet a child.

### SQUIRREL CORN—*Dicentra Canadensis* (D. C.)

This elegant species belongs to the Fumitory family, and is remarkable for its sweetness, as well as for the grace of its almost pellucid white, or pale pink, bells, and the finely dissected compound foliage of a peculiar bluish-tint of green. The Corolla is heart-shaped with slightly rounded blunt spurs, the tips of the petals projecting and rather more distinctly coloured. There is a fine variety of this flower with larger, more drooping bells, and of a decidedly pink shade.

In the rich black mould of the forest, and in rather damp situations, this species known by old settlers as Squirrel Corn and by others as Wood Hyacinth may be found. The sweet scent of the fresh flowers has suggested the last name. The round clusters of orange bulblets that are found at the base of the scape no doubt gave rise to the more common name Squirrel Corn. Whether these grain-like looking bulbs are eaten by the little ground squirrels, I do not know, the fact depends upon the authority of the Indians and old woodsmen, so we assume it is correct.

In studying the habits of this and the next species of the genus Dicentra, I have noticed some peculiarities of growth in these interesting plants which appear to have escaped the attention of the more learned botanical writers. One thing may here be mentioned, which is, the total and very rapid disappearance of the whole plant, directly the flower has perfected and ripened the seed, which is about a month after the plant has bloomed. The fine and elegantly dissected compound leaves wither away, leaving not a wreck behind to mark where the plant had grown ; delicate seedlings, indeed, may be detected near where the older plant stood, and a few golden bulblets may be found near by, under the mould, but not a vestige of the original plant remains. These golden slightly flattened bulbs are intensely bitter but not acrid or biting. I think the tiny seedlings are not the offspring of these bulbs but of the real seed—yet the bulbs will vegetate and produce living plants, as in the Tiger Lily.

All the species flourish under cultivation, and become very ornamental early border flowers ; but care should be taken to plant them in rich black vegetable mould, the native soil of their forest haunts.

This family contains another very charming species, to which the outlandish and vulgar name of " Dutchman's Breeches " has been given, and I am sorry to say has been retained in Dr. Gray's manual. A far prettier and more appropriate, because descriptive name would be that of

### FLY-FLOWER—*Dicentra Cucullaria* (D. C.)

the diverging nectaries taking just the angle of the wings of the Deer. Fly when spread for flight, and the brown tips of the four petals give the semblance of the head of the insect. The delicate pale primrose-tinted sac-like spurs of the Corolla, give a peculiar aspect to this very attractive flower, which forms one of the ornaments of the Spring. It appears early in the month of May, or, in warm and genial seasons, as early as the latter weeks in April. Like the Squirrel Corn, the foliage is finely dissected and ample ; it blooms a week earlier however.

### GOLDEN FUMITORY— *Corydalis aurea* (Willd.)

This pretty flower is also one of our native Fumitories ; it makes a good border bloomer ; is biennial in habit, seeds itself and blossoms freely. It is a low growing bushy plant, with pale bluish, finely disected foliage, and simple racemes of golden yellow flowers ; it begins to blossom very late in May, and continues all through June, and later. There is a finer, larger, more compactly growing plant, with larger flowers and foliage, found in rocky woods and islands in our backwoods' lakes. A very pretty species is *Corydalis glauca* (Pursh). This is tall and branching, with delicate flowers of bright pink, yellow and green, or white. The foliage is very blue in shade, not very abundant ; the divisions of the leaf bluish ; pods very slender, splitting and shedding bright shining seeds. It is a very pretty plant and grows readily among grasses and other wayside herbage.

### BLUE COHOSH, PAPPOOSE ROOT— *Caulophyllum thalictroides* (Michx).

Though bearing the same Indian name "Cohosh" our plant has been removed by botanists to another family, than the red and white Baneberries, or Cohoshes, which are members of the Ranunculaceæ, or Crowfoot family. There is no beauty in the blossoms of the Blue Cohosh, yet the plant is remarkable for its medicinal uses, which are well known among the Indians, and herbalists of the United States medical schools.

The round, rather large, blue berries are not the portion of the the plant that is used, but the thick knotted root-stock. The leaves are of a dull bluish green, the flowers dark purplish green, lurid in colour ; the leaves are closely folded about the thick fleshy stem when they first appear. The whole plant impresses one with the conviction that it is poisonous in its nature ; there is something that looks *uncanny* about it. Nature stamps a warning on many of our herbs by unmistakeable tokens : the glaring inharmonious colouring of some ; the rank odours exhaled by others ; the acrid biting-taste in the leaves and juices of some are safe guards if we would but heed them as warnings. The compound leafage of the Blue Cohosh breaks the ground in April, with the immature flowers- after a while the leaf spreads out, and lurid blossoms expand. The berries are set upon short thick fleshy foot-stalks, and the round hard fruit forms a loose panicle of drupe-like, naked seeds of horny texture.

The plant may be found in open woods, and grassy plain-lands—known by its large bluish green leafage, and the dark blue berries.*

RED BANEBERRY, RED COHOSH—*Actaea spicata* (*L.*) var *rubra* (Gray)

The Red Cohosh is a larger plant, with foliage coarsely veined pointed in the divisions, of a full green, sharply cleft, and toothed; flowers, white in a close tufted terminal raceme. The berries when ripe are oval, shining, of a deep red, set on slender stalks; it grows in damp rich woods.

WHITE COHOSH—*Actaea alba* (Bigel).

This is a striking looking plant when in ripe fruit, the berries are white, and shining, set on rose-red fleshy foot-stalks, the plant is branch-ing and inclined to fall prostrate from the weight of the long stalked cluster of heavy fruit. In some of its peculiar characteristics it seems to resemble the Blue Cohosh—the Indian herbalists evidently considered they were of the same nature. In none of these plants is the fruit edible.

[BELLWORT—WOOD DAFFODIL.—*Uvularia grandiflora* (Smith.)

**(PLATE II.)**

" Fair Daffodils we weep to see
Thee haste away so soon,
As yet the early rising sun
Has not attained his noon.
Stay, Stay !—
Until the hasting day
Has run,
But to the evening song ;
When having prayed together we
Will go with you along."—*Herrick*

This slender drooping flower of early spring, is known by the name of Bellwort, from its pendant lily-like bells; and by some it is better known as the Wood Daffodil, to which its yellow blossoms bear some remote resemblance.

The flowers of the Bellwort are of a pale greenish-yellow; the divisions or the petal-like sepals are six, pointed and slightly twisted or waved, the flowers droop from slender thready pedicels term-inating the branches; the stem of the plant is divided into two portions, one of which is generally barren of flowers. The leaves are of a pale green, smooth, and in the largest species, perfoliate, clasping the stem.

---

*The roots of this plant are in use with the Indian women, its common name is " Pappoose Root." Its virtues are of a singular and powerful nature, known only to the native Indian.

The root (or rhizome) is white, fleshy and tuberous.   The Bellwort is common in rich shady woods and grassy thickets, and on moist alluvial soil on the banks of streams, where it attains to the height of two feet.   It is an elegant, but not very showy flower—remarkable more for its graceful pendant straw-coloured or pale yellow blossoms, than for its brilliancy.   It belongs to a sub-order of the Lily tribe. There are three species in Canada—*Uvularia grandiflora, U. perfoliata* and *U. sessilifolia.*

### ADDER'S-TONGUE—DOG'S-TOOTH VIOLET.

*Erythronium Americanum* (Smith).

" And spot ed Adders-tongue with drooping bell,
Greeting the new-born spring."

In rich black mould, on the low banks of creeks and open wood-lands, large beds of these elegant Lilies may be seen piercing the softened ground in the month of April ; the broad lanceolate leaves are beauti-fully clouded with purple or reddish brown, and sometimes with milky white.   Each bulb of the second year's growth produces two leaves, and between these rises a round naked scape, (or flower stem), terminated by a drooping yellow bell.   The unfolded bud is striped with lines of dark purple.   A few hours of sunshine and warm wind soon expand the perianth, which is composed of six coloured recurved segments, which form a lily-like turban-shaped flower ;   each segment grooved, and spotted at the base, with oblong purplish brown dots.   The outer surfaces of three of the coloured flower leaves are marked with dark lines.   The stamens are six ; anthers oblong ; pollen of a brick-red, or dull orange color, varying to yellow.   The style is club-shaped ; stigmas three, united.

This elegant Yellow Lily bends downward when expanded, as if to hide its glories from the full glare of the sunlight.   The clouded leaves are of an oily smoothness, resisting the moisture of rain and dew.   This is one of the most elegant of our native Lilies and well worth cultivation. It blossoms early in May or late in April, and we hail it with gladness when it brightens us with a graceful golden bell at the edge of the dark forest.

The name Dog's-tooth Violet seems very inappropriate.   The pointed segments of the bell may have suggested the resemblance to the tooth of a dog ; but it is difficult to trace any analogy between this flower and the Violet, no two plants presenting greater dissimiliarty of form or habit than the Lily and the Violet, though often blended in the

verse of the poet. The American name, Adder's-tongue is more sig-
n'ificant.* This name must refer to the red pointed anthers rather than
the foliage, as some have suggested.

The White Flowered Adder's-tongue, *Erythronium albidum* (Nutt),
grows in the more western portions of Canada, as on the shores of Lake
Huron.

WHITE TRILLIUM—EASTER FLOWER—*Trillium grandiflorum* (Salisb).

> " And spotless lilies bend the head
> Low to the passing gale. "

Nature has scattered with no niggardly hand, these remarkable
flowers, over hill and dale, wide shrubby plain and shady forest glen.
In deep ravines, on rocky islets, the bright snow white blossoms of the
Trilliums greet the eye and court the hand to pluck them. The old
people in this part of the Province call them by the familiar name of
Lily. Thus we have Asphodel Lilies, Douro Lilies, &c. In Nova
Scotia they are called Moose-flowers, probably from being abundant in
the haunts of Moose-deer. In some of the New England States the
Trilliums, white and red, are known as the " Death-flower," but of the
origin of so ominous a name we have no record. We might imagine it
to have originated in the use of the flower to deck the coffin or graves
of the dead. The pure white blossoms might serve not inappropriately
for emblems of innocence and purity, when laid upon the breast of the
early dead. The darker and more sanguine hue of the red species,
might have been selected for such as fell by violence ; but these
are but conjecture. A prettier name has been given to the Nodding
Trillium (*T. cernuum*) : that of " Smiling Wake-robin," which seems to
be associated with the coming of the cheerful chorister of early spring,
" The household bird with the red stomacher," as Bishop Carey †
calls the Robin Red-breast. The botanical name of the Trillium
is derived from *trilex*, triple, all the parts of the plant being in
threes. Thus we see the round fleshy scape furnished with three
large sad green leaves, two or three inches below the flower, which is
composed of a calyx of three sepals, a corolla of three large snow white,
or, else, chocolate red petals : the styles or stigmas, three ; ovary three
celled ; and the stamens six, (which is a multiple of three.) The white
fleshy tuberous root is much used by the American Schools of Medicine in
various diseases, also by the Indian herb doctors.

*Trillium grandiflorum* is the largest and most showy of the white
species. *Trillium nivale* or Lesser Snowy Trillium is the smallest ;

---

* The name Dog's-tooth refers to the shape of the small pointed white bulbs of the common European species, so well known in English gardens.--PROF. LAWSON.

† An old writer in the time of James I., and tutor to one of the daughters of Charles I.

this last blooms early in May.    May and June are the months in which
these flowers appear.    The white flowered Trilliums are subject to many
varieties, and accidental alterations.    The green of the sepals is often
transferred to the white petals in *T. nivale;* some are found handsomely
striped with red and green, and in others the very foot-stalks of the
almost sessile leaves are lengthened into long petioles.    The large
White Trillium is changed, previous to its fading, to a dull reddish lilac.

PURPLE TRILLIUM—BIRTH-ROOT—*Trillium erectum.*    (Lin.)

> " Bring flowers, bring flowers o'er the bier to shed
> A crown for the brow of the early dead.
> Though they smile in vain for what once was ours,
> They are love's last gift, bring flowers, bring flowers."—*Hemans.*

Gray and other botanical writers call this striking flower " Purple
Trillium ;" it should rather be called red, its hue being decidedly more
red than purple ; and in the New England States it is called by the
country folks, The Red Death-flower, in contrast to the larger White
Trillium or White Death-flower.    *T. erectum* is widely spread over
the whole of Canada.    It appears in the middle of May, and continues
blooming till June, preferring the soil of damp, shady woods and thickets ;
but it takes very kindly to a shaded border in the garden, where it
increases in size, and becomes an ornamental spring flower.

" Few of our indigenous plants surpass the Trillium in elegance
and beauty, and they are all endowed with valuable medicinal properties.
The root of the Purple Trillium is generally believed to be the most
active.    Tannin and Bitter Extract form two of its most remarkable
ingredients."    So says that intelligent writer on the medicinal plants of
North America, Dr. Charles Lee.

The Red Trilliums are rich but sombre in colour, the petals are
longish-ovate, regular, not waved, and the pollen is of a greyish dusty
hue, while that of the White species is bright orange-yellow.    The leaves
are of a dark lurid green, the colouring matter of the petals seems to
pervade the leaves ; and here, let me observe that the same remark may
be made of many other plants.    In purple flowers we often perceive the
violet hue to be perceptible in the stalk and under part of the leaves,
and sometimes in the veins and roots.    Red flowers, again, show the
same tendency in stalk and veins.    Where the flower is white the leaves
and veinings, with the stem and branches, are for the most part of a
lighter green, more inclining to the yellow or else bluish tinge of green.

The Blood-root in its early stage of growth shews the Orange juice
in the stem and leaves, as also does the Canadian Balsam, and many others
that a little observation will point out.    The colouring matter of flowers
has always been, more or less, a mystery to us : that light is one of the

great agents can hardly for a moment be doubted, but something also may depend upon the peculiar quality of the juices that fill the tissues of the flower, and on the cellular tissue itself. Flowers deprived of light we know are pallid and often colourless, but how do we account for the deep crimson of the Beet-root, the rose red of the Radish, the orange of the Rhubarb and Carrot, which roots, being buried in the earth, are not subject to the solar rays? The natural supposition would be that all roots hidden from the light would be white, but this is by no means the case. The question is one of much interest and deserves the attention of all naturalists, and especially of the botanical student.

What shall we say to the rich colour of the Ruby, Carbuncle, Amethyst, Topaz and Emerald, taken from the darkness of the mine; can it be that all are really colourless till the light is admitted to them, and the different conditions of the crystallised forms catch, imprison and forever hold fast the glorious rays of light.

### PAINTED TRILLIUM—*Trillium erythrocarpum*, (Mx.)
#### (PLATE III.)

This beautifully ornamental species is of rare occurrence in our woods. The flower is elegantly tinged with soft pink veinings on the white, waved, and pointed, petals; the base of each is richly coloured and shaded from deep red to pale rose, which colour indeed is slightly diffused through the flower. Leaves distinctly petioled, broad at the base, waved at the margins and sharply pointed. The whole plant, from six to nine inches in height. The specimen from which the drawing is taken was found in May, near Ottawa, where it is not uncommon. The under-surface of the leaves is slightly tinged with purple.

Though scarce in our western woods, Gray says the Painted Trillium may be found as far northward as Lake Superior and New England, and also southward in the Alleghanies and Virginia.

### ROCK COLUMBINE—*Aquilegia Canadensis* (Lin).

"The graceful Columbine all blushing red,
Bends to the earth her crown
Of honey-laden bells."

This graceful flower enlivens us all through the months of May and June by its brilliant blossoms of deep red and golden yellow.

In general outline the Wild Columbine resembles its cultivated sisters of the garden, but is more light and airy in habit. The plant throws up many tall slender stalks, furnished with leafy bracts, from which spring other light stems terminated by little pedicels, each bearing a large drooping flower and bud, which open in succession.

The flower consists of five red sepals and five red petals; the latter are hollowed, trumpet-like at the mouth, ascending; they form narrow tubes, which are terminated by little round knobs filled with honey. The delicate thready pedicel on which the blossom hangs causes it to droop down and thus throw up the honey-bearing tubes of the petals; the little balls forming a pretty sort of floral coronet at the junction with the stalk.

The unequal and clustered stamens, and five thready sty'es of the pistil, project beyond the hollow mouths of the petals, like an elegant golden fringed tassel ; the edges and interior of the petals are also of a bright golden yellow. These gay colors are well contrasted with the deep green of the root leaves and bracts of the flower stalks. The bracts are lobed in two or three divisions. The larger leaves are placed on long foot stalks, each leaf is divided into three leaflets, which are again twice or thrice lobed, and unequally notched; the upper surface is smooth and of a dark rich green, the under pale and whitish. As the flowers fade the husky hollow seed pods become erect.

The wild Columbine is perennial and very easily cultivated. Its blossoms are eagerly sought out by the Bees and Humming-birds. On sunny days you may be sure to see the latter hovering over the bright drooping bells, extracting the rich nectar with which they are so bountifully supplied. Those who care for Bees, and love Humming-birds, should plant the graceful red-flowered Columbine in their garden borders. Indeed this elegant ornamental species should find a place in every garden. I have seen a striking effect produced by a number of these flowers grown together.

In its wild state it is often found growing among rocks and surface stones, where it insinuates its roots into the clefts and hollows that are filled with rich vegetable mould; and thus, being often seen adorning the sterile rocks with its bright crown of waving blossoms, it has obtained the name of Rock Columbine.

PAINTED CUP—SCARLET CUP.—*Castilleia coccinea.* (Spreng.)

> Scarlet tufts
> Are glowing in the green, like flakes of fire ;
> The wanderers of the prairie know them well,
> And call that brilliant flower the Painted Cup.—*Bryant,*

This splendidly-coloured plant is the glory and ornament of the plain-lands of Canada. The whole plant is a glow of scarlet, varying from pale flame-colour to the most vivid vermillion, rivalling in brilliancy of hue the Scarlet Geranium of our gardens.

The Painted Cup owes its gay appearance, not to its flowers, which are not very conspicuous at a distance, but to the deeply-cut, leafy bracts that enclose them and clothe the stalks, forming at the ends of the flower-branches clustered rosettes.

The flower is a flattened tube, bordered with bright red, and edged with golden yellow. Stamens, four ; pistil, one, projecting beyond the tube of the calyx ; the capsule is many-seeded. The radical or root leaves are of a dull, hoary green, tinged with reddish purple, as also is the stem, which is rough, hairy, and angled. The bracts, or leafy ap. pendages which appear on the lower part of the stalk, are but slightly tinged with scarlet, but the colour deepens and brightens towards the middle and summit of the branched stem.

The Scarlet Cup appears in May, along with the white and red Trilliums ; but these early plants are small ; the stem simple, rarely branched, and the colour of a deeper red. As the summer advances, our gallant, soldier-like plant, puts on all its bravery of attire. All through the glowing harvest months, the open grassy plains and the borders of the cultivated fields are enriched by its glorious colours. In favourable soil the plant attains a height of from 2 to 3 feet, throwing out many side branches, terminated by the clustered, brilliantly-tinted bracts ; some heads being as large as a medium-sized rose. They have been gathered in the corners of the stubble fields on the cultivated plains, as late as October. A not uncommon slender variety occurs of a pale buff, and also of a bright lemon colour. The American botanists speak of *Castilleia coccinea* as being addicted to a low, wettish soil, but this has not been my experience ; if you would find it in its greatest perfection, you must seek it on the high, dry, rolling plains of Rice-lake, Brantford, the Humber, to the north of Toronto, Stoney Lake, the neighborhood of Peterboro, and similar localities.

For soil, the Scarlet Cup seems to prefer light loam, and evidently courts the sunshine rather than the shade. If it could be prevailed upon to flourish in our garden borders, it would be a great acquisition, from its long continuance in flower, and its brilliant colouring. The seed is light brown, contained in thinnish capsules, ripe in September. Gray says : " Herbs parasitic on roots," but our brave plant is no parasite but grows freely on open ground. Neither is it found with us in low wettish places ; it loves the light and would not flourish in shade. It is essentially a " Prairie flower." I have had bright specimens from our North-West, and also from Wisconsin and Dakota, U. S.

These lovely plants, like many others that adorn our Canadian woods and wilds, yearly disappear from our midst, and soon we shall seek them, but not find them.

We might say with the poet :

> "'T was pity nature brought ye forth
>   Merely to show your worth,
>     And lose ye quite !
> But ye have lovely leaves, where we
>   May read how soon things have
>     Their end, though ne'er so brave ;
> And after they have shown their pride,
>   Like you awhile they glide
>     Into the grave."—*Herrick.*

I do not know if our brave Scarlet Cup, of Canada, has any flora. relationship to an herb known in the Old Country as " Clary " or by its local and descriptive name of " Eye-bright." It is an old-fashioned flower, sometimes found in cottage gardens. I remember its curious coloured leaves and bracts attracted my notice where first I saw it, in a neglected corner of a poor old woman's garden. There were two varieties, one with the dull, veiny leaves, bordered with purple, as if the leaves had been dipped into some logwood dye, the other with a full pink. I forget, in the long lapse of time since I saw the plants, if the flower itself was pretty, or partook of the same tint of colour as the foliage, but the great marvel consisted in the black, oval seeds, not very large, about the size of the seed of the Sage. This wonderful seed, Nannie Prime told me, gave the name to the plant " Eye-bright," though, she added, " the learned gardener folk do call it ' Clary.' If any dust or motes, or any bad humors, are in the eye and one of these seeds be put into the corner of the eye, it will gather it all round itself and clear the precious sight ; and this is why folks do give it the name of ' Eye-bright.' Sure, Miss, the Lord gave this little seed for a cure for us poor folk, and no doubt the whole plant is good for other complaints, as many of our *harbs* be, if we did but use them right." We know of no especial healing virtue contained in the seed or leaves of our beautiful Scarlet Cup ; but it charms the eye and delights us, and that is God's gift also. There seems to be no actual void, no space unfilled in God's creation. Something fills up all vacancies, either in vegetable or animal life ; unseen organisms, too subtle and too fine to become visible to our unassisted vision, have their existence though we behold them not.

> " Father of earth and heaven, all, all are thine ;
>   The boundless tribes in ocean, air and plain,
>   And nothing lives, and moves, and breaths in vain.
> Thou art their soul, the impulse is divine :
> Nature lifts loud to Thee her happy voice,
>   And calls her caverns to resound Thy praise ;
>   Thy name is heard amid her pathless ways,
> And e'en her senseless things in Thee rejoice."

PLATE II.

WILD GINGER—-*Asarum Canadense* (L.)

This is a singular herbaceous plant, chiefly found in bush-wood and damp, rich meadow-land. The leaves are wide, rounded kidney-form with deep sinuses. The flower, on a short peduncle, springs from the root-stock and appears below the leaves close to the ground, seldom more than one to each plant; it is campanulate with sharp pointed segments of a deep chocolate colour. The floral envelope consists of a calyx, but no corolla; the creeping, thick fleshy root-stock is warm, pungent and aromatic. It is a coarse singular looking plant much used in Indian medicine craft.

SHOWY ORCHIS.—*Orchis spectabilis*, (L.)

" Full many a gem of purest ray serene,
The dark unfathomed caves of ocean bear ;
Full many a flower is born to blush unseen,
And waste its sweetness on the desert air."—*Gray.*

Deep hidden in the damp recesses of the leafy woods, many a rare and precious flower of the Orchis family blooms, flourishes, and decays, unseen by human eye, unsought by human hand, until some curious, flower-loving botanist plunges amid the rank, tangled vegetation, and brings its beauties to the light. One of these lovely natives of our Canadian forests is known as *Orchis spectabilis*—Beautiful Orchis— or Showy Orchis. This pretty plant is not, indeed, of very rare occurrence ; its locality is rich maple and beechen woods all through Canada. The colour of the flower is white, shaded, and spotted with pink or purplish lilac ; the corolla is what is termed ringent or gaping, the upper petals and sepals arching over the waved lower-lipped petal. The scape is smooth and fleshy, terminating in a loosely-flowered and many-bracted spike ; the bracts are dark-green, sharp-pointed, and leafy ; the root a bundle of round white fibres ; the leaves, two in number, are large, blunt, oblong, shining, smooth, and oily, from three to five inches long, one larger and more pointed than the other. The flowering time of the species is May and June. The exquisite cellular tissues of many of our flowers of this order delight the eye, and give an appearance of great delicacy and grace to the blossoms. In this charming species the contrast between the lilac purple colour of the arching petals and sepals, and the almost pellucid lower lip or somewhat broadly lobed under petal, is very charming. The large shining leaves lie close to the ground when the plant is in flower. Transplanted to gardens, the Showy Orchis rarely survives the second season of removal from the forest shade. It will not grow freely, exposed to cold wind, or glaring sunlight. It loves moist heat ; the conservatory would probably suit it, and it would be worth a trial there.

c

### Lady's Slippers—Moccasin Flowers.

Among the many rare and beautiful flowers that adorn our native woods and wilds, few, if any, can compare with the lovely plants belonging to the Orchis family. Where all are so worthy of notice it was difficult to make a choice; happily there is no rivalry to contend with in the case of our Artist's preferences. We will, however, first treat of the Cypripediums or Lady's Slippers, better known by the name of Moccasin-Flowers, a name common in this country to all the species. The plants of this family are remarkable, alike for the singular beauty of their flowers, and the peculiar arrangement of the internal organs. In the Linnæan classification they were included, in common with all the Orchids, among the Gynandria.

Whether we regard these charming flowers for the singularity of their form, the exquisite texture of their tissues, or the delicate blending of their colours, we must acknowledge them to be altogether lovely and worthy of our admiration.

One of the rarest, and at the same time most beautiful and curious is the

RAM'S-HEAD ORCHIS—*Cypripedium arietinum* (R-Br.)
(PLATE VII.)

which has smooth glaucous green leaves, and small purplish flowers bearing a close resemblance to a ram's head with the horns and ears and a tuft of wool on the top of the head. It is seldom over six inches in height, and grows in cold peat bogs, and flowers in July; associated with it we find our most gorgeous representative of the family, the

SHOWY LADY'S SLIPPER OR PINK FLOWERED MOCCASIN PLANT—
*Cypripedium spectabile* (Swartz).

It grows chiefly in tamarack swamps, and near forest creeks, where, in groups of several stems, it appears, showing its pure blossoms among the rank and coarser herbage. The stem rises to the height of from 18 inches to two feet. The leaves, which are large, ovate, many nerved and plaited, sheathing at the base, clothe the fleshy stem, which terminates in a single sharp-pointed bract above the flower. The flowers are terminal and generally solitary, although old and strong plants will occasionally bear two or even three blossoms on one stem. The unfolded buds of this species are most beautiful, having the appearance of slightly flattened globes of delicately-tinted rice-paper.

The large sac-like inflated lip is slightly depressed in front, tinged with rosy pink, and striped. The pale thin petals and sepals, two of each, are whitish at first, but turn brown when the flower is more advanced towards maturity. The sepals may be distinguished from the petals ; the former being longer than the latter, and by being united at the back of the flower. The column on which the stamens are placed is three-lobed ; the two anthers are placed one on either side, under the two lobes ; the central lobe is sterile, thick, fleshy, and bent down, somewhat blunt and heart-shaped. The root of the Lady's Slipper is a bundle of white fleshy fibres.

One of the remarkable characteristics of the flowers of this genus, and of many of the natural order to which it belongs, is the singular resemblance the organs of the blossom bear to the face of some animal or insect. Thus the face of an Indian hound may be seen in the Golden-flowered *Cypripedium pubescens;* that of a sheep or ram, with the horns and ears, in *C. arietinum;* while our " Showy Lady's Slipper " displays the curious face and peering black eyes of an ape.

A rarer species is the

STEMLESS LADY'S SLIPPER—*Cypripedium acaule* (Ait.)

It differs from the former species by the sac, which is large and of a beautiful rose tint exquisitely veined with deeper red zig-zag lines, not being closed; but merely folded over in front; this is not observable until you examine it closely. The scape rises from between the two large oval leaves which lie horizontally on the mosses amidst which the plant grows.

A time will come when these rare productions of our soil will disappear from among us, and will be found only in those waste and desolate places, where the foot of civilized man can hardly penetrate ; where the flowers of the wilderness flourish, bloom and decay unseen but by the all-seeing eye of Him who adorns the lonely places of the earth, filling them with beauty and fragrance.

For whom are these solitary objects of beauty reserved ? Shall we say with Milton :—

> " Thousands of unseen beings walk this earth,
> Both while we wake and while we sleep :—
> And think though man were none,—
> That earth would want spectators—God want praise.

YELLOW LADY'S SLIPPERS—*Cypripedium parviflorum* (Salisb.) and *Cypripedium pubescens* (Willd.)

" And golden slippers meet for Fairies' feet."

Of the golden-flowered Moccasin flowers, we boast of two very beau tiful species, *C. pubescens,* Hairy Moccasin flower, and *C. parviflorum* " Lesser-flowered Moccasin flower." The larger plant is the more showy ; the smaller the more graceful, and has a delicate fragrance which is not so strong in the larger flower. The long spirally twisted petals and sepals of a purplish brown colour, sometimes tinted and veined with red, give this smaller flower a very elegant appearance, though the rich golden hue of the larger is more striking to the eye.

*C. parviflorum* affects the moist soil of wet grassy meadows and swamps, while the larger plant loves the open plain lands among shrubs and tall grasses. In the month of June when it may be seen beside the gay Painted Cup (*Castilleia coccinea*), the Blue Lupine (*L. perennis*), the larger White Trillium, and other lovely wild flowers, it forms a charming contrast to their various colours and no less varied forms.

The stem of the larger Moccasin flower is thick and leafy, each many-nerved leaf sheathing the flowers before they open. The flowers are from one to three in number, bent forward, drooping gracefully downwards. The golden sac-like lip is elegantly striped and spotted with ruby red ; the twisted narrow petals, and sepals, two in number of each kind, are of a pale fawn colour, sometimes veined and lined with a deeper shade.

### Wild Garlic—Wild Leek.—*Allium tricoccum* (Ait.)

As soon as the warm rays of early spring sunbeams have melted the snow in the woods, we see the bright, closely-folded and pointed leaves of the Wild Garlic, or Wild Leek as it is commonly called, piercing through the carpet of dead leaves that thickly covers over the rich black mould, the refuse of many years of former decayed foliage. The cattle, that have been for many months deprived of green food, eagerly avail themselves of the first appearance of the succulent and welcome leaves of the Garlic. The milk of the cows becomes so strongly flavoured with the disagreeable odour of the oily vegetable that the milk and butter are rejected, and can only be used by persons who are indifferent to the nature of their food, caring more for quantity than quality ; but the generality of people turn away with a feeling of disgust from leeky butter and leeky milk. It is, however, a consolation to the thrifty farmer to know that, like many other evils, it has its palliative. The cows and oxen that have been brought low in flesh and strength during the long, hard winter, are speedily restored to health by feasting upon this otherwise objectionable food.

It is a pleasant plant to the eye—the rich verdure of the broad succulent leaves springing so freshly where all was barren and unsightly —and later in the season, the tall heads of pretty, pale blossoms are not without attraction, though not nice to place in a bouquet of sweeter flowers.

Before so many extensive tracts of forest had been cut down, the Wild Garlic was to be found in all beech and maple woods. But it is becoming very rare, and you hear no more complaints of leeky milk and butter.

### PHLOX—*Phlox divaricata* (L.)

We have in Canada several species of this family, and all are worthy of cultivation. *Phlox divaricata* is found on dry grassy wastes by forest roads, in shady spots. It is a plant of slender growth, about twelve or eighteen inches high, with slender lanceolate pointed leaves somewhat clasping the stem; flowers in a flat spreading head terminal on the slightly stalked branches, corolla salver shaped, primrose-like; calyx with slender pointed sepals; colour of the petals, pale lilac, scalloped at the edges—it is an elegant species. A small variety of this beautiful flower has also been found in low meadows near the Ottawa river growing in great profusion in some of the North-eastern townships—its beautiful blue flowers formed an attractive feature in the landscape.

A gentlemen who had an especial love for the beauties of nature was much struck with the beauty of this very lovely flower, and brought home some roots; the plant was then in full bloom. They continued to flourish till the following spring, when they disappeared entirely. The leaves were of a full rich glossy green, delicately fringed with silky purplish hairs; flowers, not so large as the *P. divaracata* found here; heads loose on long footstalks springing from between the slightly clasping leaves; roots white, fibrous.

A charming little dwart Phlox is that known by the gardener as Moss Pink, or Lake Erie Moss. The slender pointed grassy looking foliage and abundant pink flowers, its low tufted growth and hardy character, make it most valuable as an edging for flower beds. It comes early and remains for some time in bloom, and even when the blossoms have faded, the bright cheerful verdure that remains, has a good effect as a pretty edging to the beds. It grows in large cushion-like plots when not used as an edging for borders.

### GOLD THREAD—*Coptis trifolia* (Salisb).

In the deep shady forest we are attracted by the bright glossy thrice parted (trifoliate) leaves of this pretty plant. In early Spring its delicate white starry flowers, on upright slender foot-stalks appear, just peeping above the mosses among which it delights to grow. The modest pearly white star-shaped blossoms, contrast well with the dark evergreen shining leaves, and orange thready rootlets, that may be seen among the light feathery mosses, hardly concealed, for they are barely covered by the mould in which they grow. The orange fibrous roots and rootlets are intensely bitter, and are much used by the old settlers as tonic remedies against weakness in children when brought low by fever and ague : more especially is it used as a wash for sore ulcerated mouths, as thrush in young infants. The Indian women use it for their little ones in case of sore mouth and sore gums in teething. I once saw the small evergreen leaves of the Gold Thread applied to a very different purpose —that of trimming evening dresses of clear white muslin, and as the heat of the room had little effect on them they looked fresh and singularly ornamental on the young ladies that had so tastefully arranged the leaves on their simple white dresses.

I have noticed the term " Gold-thread " applied lately to one of the species of Dodder, that singular parasite, but it was by a person apparently unacquainted with our elegant little forest evergreen *Coptis trifolia.*

### BUNCH BERRY—SQUAW BERRY—*Cornus Canadensis* (Lin.)

This elegant and attractive little plant is met with most commonly in beds, beneath the shade of evergreens, Hemlocks and Spruces, it multiplies by its creeping root-stock as well as by the drupe-like berry. Its popular name in the back-woods, is the Squaw-berry, and also Bunch-berry. It is a truly lovely little plant—a perfect forest gem.

In height our tiny Dogwood rarely exceeds four or six inches ; the stem is leafy, the upper leaves form a whorl round the flowers, which are enclosed by the white corolla-like involucre, which is more conspicuous than the tiny terminal umbel of little flowers with their dark anthers. The flowers are succeeded by small round berries which become brilliantly scarlet by the end of the summer, appearing like a bright red coral ring surrounded by the whorl of dark green, somewhat pointed, veiny leaves.

From its love of shady damp soil, this little plant would grow under cultivation, if suitable localities were selected in shrubberies, among evergreens and in rock-work not much exposed to the sun. This low Cornel is very ornamental, both in flower and fruit. The berries are sweet but insipid. The Indian women and children eat them and say, " good to eat for Indian." The taste of the Indian is so simple and uncultivated that they will eat any fruit or vegetable that is innoxious, apparently indifferent to its flavour.

The poor squaw gathers her handful of berries, and goes her way contented with her forest fare, from which the more luxurious children of civilization would turn away with contempt, or admire their beauty possibly, and then cast them away as worthless. Few indeed think of the lessons that may be learned even from the humblest forest flower, speaking to their hearts of the loving care of the great Creator, who provideth alike for all his creatures; the wild berry to feed the wild bird, the Squirrel and Field-mouse. He openeth His hand and filleth all things living with plenteousness.

There are, among other species of the Dogwood family that might be enumerated as indigenous to this Western part of Canada, some with blue berries, some with white, some with red and others with dark steel coloured fruit. The dwarf Cornel, *C. Canadensis* is the smallest species, the rough, bushy round-leaved *C. circinata* the second; *C. floridus* the largest: all are tonics, and bitter; some are used in medicine; others in dyeing by the natives. The berries of several species are largely sought for and form food for the wild ducks that haunt the borders of marshes and lake shores where these shrubs abound.

The Cornel seems to have a wide geographical range, it being found not only in the Eastern States of N. America, but in the colder parts of Canada, westerly and northerly, and extends even to the borders of the Arctic Zone. I have before me a specimen of a closely allied species from North Cape, Norway, which was gathered by a friend among the dark evergreen glades of that far-off land. The tiny plant is smaller, and has a more pinched and starved look than our more vigourous plant, otherwise there is no apparent difference. The early frosts of Autumn give a pretty purple shade to the surface of the leaves of our little forest Dogwood, but they do not wither, remaining fresh and persistent through the winter beneath the snow.

#### TWISTED STALK—*Streptopus roseus* (Mx.)

This is a graceful plant with pretty pink, striped, bells belonging to the Lily family. We find it in the forest as well as in open grassy thickets. The stalk is divided into two or three branches, bearing on the underside several pairs of graceful, pendant bells on thready, twisted, foot-stalks. The tips of the segments are pointed and slightly recurved. The berries are red, round and seeded with several hard, bony nutlets. The flower is scentless. The foliage is of a light yellowish green, many nerved, oval and pointed. Associated with this there often may be found in the deep shade of pine woods, as well as in the rich black leaf mould of the hardwood forest, The False Solomon's Seal (*Polygonatum biflorum, L.,*) which has pale greenish-tinged bells and large blue berries. The leaves are of a dark bluish green. The stem is simple and bends gracefully. The flowers, notwithstanding the name, are mostly solitary. Our woods hide within their shades many a lovely flower, seen only by the Indian hunter and the backwoods lumberer or the axe-man; by the former they are noted for some medicinal or healing quality, by the latter they are trodden under foot, while to the uneducated settler whose business is to clear the forest land of the trees and wild productions of the soil, on which the life supporting grain and roots are to be sown or planted, these natural beauties have no value or charm, and he says " Cut them down, why cumber they the ground." In these things he sees not the works of the Creator ; they are, in his eyes, " weeds—weeds—weeds, nothing but weeds."

Our Bellworts and Trilliums, Smilacinas and Orchids are among our most interesting and attractive native forest flowers, but as the woods are levelled and the soil changed, by exposure to the influence of the elements and the introduction of foreign plants, our native vegetation disappears, and soon the eye that saw and marked their lovely forms and colours will see them no more.

### MAY-APPLE—MANDRAKE—*Podophyllum peltatum* (L.)

The Mandrake or May-Apple is chiefly found in the rich black soil of the forest, where partially clear of underwood ; in such localities it forms extensive beds. When the broad umbrella-like leaf first breaks the soil, early in May, it comes up closely folded round the simple fleshy stem, in colour of a deep bronze or coppery hue, smooth and shining, but assumes a lighter shade of green as it expands. The blossom appears first as a large round green bud between the axils of the two broad peltate, lobed and pointed leaves ; the first year's leaves are single and smaller and the young plant is flowerless.

The corolla of the flower consists of from six to nine concave greenish-white thick petals ; sepals (or calyx leaves) six ; the edges of the petals are generally torn or ragged, the handsome flower slightly drooping between the two large leaves gives out a powerful scent—not agreeable if inhaled too closely, but pleasant at a little distance.

The plant increases by buds from the thickly matted fleshy root-stock ; the roots form a singular net-work under the soft vegetable mould, spreading horizontally, at every articulation sending up a pair of fruit-bearing scapes. The single-leafed plant is most probably a seed-ling of the former year.

The fruit of the May-Apple is a large fleshy berry ; the outer rind is, when ripe, yellow, otherwise darkish-green and of a rank unpleasant flavour ; the inner or pulpy part is white, soft and filled with somewhat bony light-brown seeds. When not over-ripe this pulpy part may be eaten ; it is sub-acid and pleasant. The fruit makes a fine preserve with white sugar and when flavoured with lemon-peel and ginger ; but the outer coat I would not make use of. The fruit is ripe in August, but should be gathered, when the first yellow spots on the outer coat indicate ripeness, and laid in a sunny window for a few days.

The medicinal value of the root of this remarkable plant is now so well established that it has superseded the use of Calomel in complaints of the liver with most medical practitioners in this country, but so powerful are its properties that it should never be used by unskilful persons. Ignorant persons have been poisoned by mistaking the leaves for those of the Marsh Marigold *(Caltha palustris)* and using them as a pot herb. A case of this kind occurred some years ago, whereby several persons were poisoned. At that time there was no attempt made by the backwoods settlers to cultivate vegetables, and they made use of many of the wild herbs with very little knowledge of their sanative or injurious qualities.

### American Brooklime—*Veronica Americana* (Schw.)

" Flowers spring up and die ungathered."—*Bryant*

In the language of flowers the blossoms of the Veronica or Speedwell are said to mean undying love, or constancy, but the blossoms of the Speedwell are fugacious, falling quickly, and therefore, one would say, not a good emblem of the endurance of love or friendship.

Sweet simple flowers are the wild Veronicas, chiefly inhabiting damp overflowed ground, the borders of weedy ponds and brooks, whence the names of Brooklime and Marsh Speedwell, Water Speedwell, and the like. Some of the species are indeed found mostly growing on dry hills and grassy banks, cheering the eye of the passing traveller by their slender spikes of azure flowers. This species is often known by the pretty name of Forget-me-not, though it is not the true " Forget-me-not," which is *Myosotis palustris*, also with the rest of its family called " Scorpion-grass" ; from the small buds, before expansion, having the petals twisted and forming a small coil at the tips of the branches. The American Brooklime is one of the prettiest of the native Veronicas, and may easily be recognized by its branching spikes of blue flowers, and veiny, partially heart-shaped leaves. It is but little that we have to say of our pretty native wildling, for its delicacy and harmless qualities are all that require notice about it. The traveller passes it by with scarcely a commendatory glance ; its fleeting pale blue, scentless blossoms, which fall at a touch, scarcely attract the little children when gathering flowers by the wayside brooks. It remains with the true lover of flowers, even if they be only homely weeds, to examine and appreciate the inimitable beauty and wisdom shown in their several parts, each so wisely fitted to perform its part according to the Divine Maker's Will.

### Wood Geranium—*Geranium maculatum* (L.)

There are but few flowers of the Cranes-bill family in Canada. The one most worthy of notice is the Wood Geranium. This is a very ornamental plant ; its favourite locality is in open. grassy thickets, among low bushes, especially those tracts of country known as Oak-openings, where it often reaches to the height of from two to three feet, throwing out many branches, adorned with deep lilac flowers ; the half-opened buds are very lovely. The blossom consists of five petals, obtuse, and slightly indented on their upper margins, and is lined and delicately veined with purple. The calyx consists of five pointed sepals : stamens ten ; the anthers are of a reddish brown ; styles five, cohering at the top. When the seed is mature these curl up, bearing the ripe brown seed adhering to the base of each one. The common name, Cranes-bill, has been derived from the long grooved and stork-like beak composed of the styles. The Greek name of the plant means a Crane. The whole plant is more or less beset with silvery hairs. The leaves are divided into about five principal segments ; these again are lobed and cut into sharply pointed, irregularly sized teeth. The larger hairy root leaves are often discoloured with red and purplish blotches, whence the specific name (*maculatum*), spotted, has been given to this species.

The flower stem is much branched, and furnished with leafy bracts ; the principal flowers are on long stalks, usually three springing from a central branch and again subdividing into smaller branchlets, terminating in buds, mostly in threes, on drooping slender pedicels ; as the older and larger blossoms fall off a fresh succession appears on the side branches, furnishing rather smaller but equally beautiful flowers. Gray gives the blooming season of the Cranes-bill from April to July, but with us it rarely appears before June, and may be seen all through July and August. Besides being very ornamental, our plant possesses virtues which are well known to the herbalist as powerful astringents, which quality has obtained for it the name of *Alum-root* among the country people, who use a decoction of the root as a styptic for wounds ; and sweetened, as a gargle for sore-throat and ulcerated mouth ; it is also given to young children to correct a lax state of the system. Thus our plant is remarkable for its usefulness as well as for its beauty. A low growing showy species, with large rose-coloured flowers and much dissected leaves, may be found on some of the rocky islets in Stoney

Lake, Ont. The slender flower stem is about six inches in height, springing from a leafy involucre, which is cut and divided into many long and narrow segments; flowers generally from one to three, terminal on the little bracted foot-stalks. The seed vessels not so long as in the Wood Geranium.

Besides the above named we have some smaller species. The well known Herb Robert (*G. Robertianum*, (L.) which is said to have been introduced from Britain; but it is by no means uncommon in Canada, in half cleared woodlands and by waysides, attracting the eye by its bright pink flowers, and elegantly cut leaves, which become bright red in the fall of the year. This pretty species is renowned for its rank and disagreeable odour, and so it is generally passed by as a weed in spite of its very pretty bright pink blossoms.

Another small-flowered species, with pale insignificant blossoms is also common as a weed by road sides and in open woods, this is *G. pusillum*, smaller Cranes-bill; it also resembles the British plant, but is of too frequent occurrence in remote localities to lead us to suppose it to be otherwise than a native production of the soil; we find it often in very remote places in our forest clearings and road-side wastes.

CHICKWEED WINTERGREEN—*Trientalis Americana* (Pursh).

This pretty starry-flowered little plant is remarkable for the occurrence of the number seven in its several parts, and was for some time cherished by botanists of the old school as the representative of the class Heptandria.

The calyx is seven parted; the divisions of the delicate white corolla also seven; and the stamens seven. The leaves form a whorl at the upper part of the stem, mostly from five to seven, or eight; the leaves are narrow, tapering at both ends, of a delicate light-green, thin in texture, and of a pleasant sub-acid flavour. The star-shaped flowers, few in number, on thread-like stalks, rise from the centre of the whorl of leaves, which thus forms an involucre to the pretty delicate starry flowers. This little plant is frequently found at the roots of trees; it is fond of shade, and in light vegetable mould forms considerable beds; the roots are white, slender and fibrous; it is one of our early May flowers, though, unless the month be warm and genial, will delay its opening somewhat later. In old times, when the herbalists gave all kinds of fanciful names to the wild plants, they would have bestowed such a name as "Herbe Innocence" upon our modest little forest flower.

LARGE BLUE FLAG—FLEUR–DE–LUCE—*Iris versicolor* (L.).

> Lilies of all kinds,
> The fleur-de-luce being one. — *Winter's Tale.*

This beautiful flower abounds all through Canada, and forms one of the ornaments of our low, sandy flats, marshy meadows and overflowed lake shores ; it delights in wet, muddy soil, and often forms large clumps of verdure in half-dried ponds and similar localities. Early in spring, as soon as the sun has warmed the waters, after the melting of the ice, the sharp sword-shaped leaves escaping from the sheltering sheath that enfolded them pierce the moist ground, and appear, forming beds of brilliant verdure, concealing the swampy soil and pools of stagnant water below. Late in the month of June the bursting buds of rich purple begin to unfold, peeping through the spathe that envelopes them. A few days of sunshine, and the graceful petals, so soft and silken in texture, so variable in shades of colour, unfold : the three outer ones reflexed, droop gracefully downwards, while the three innermost, which are of paler tint, sharper and stiffer, stand erect and conceal the stamens and petal-like stigmas, which lie behind them : an arrangement so suitable for the preservation of the fructifying organs of the flower, that we cannot fail to behold in it the wisdom of the great Creator. The structure of the cellular tissue in most water plants, and the smooth, oily surface of their leaves, has also been provided as a means of throwing off the moisture to which their place of growth must necessarily expose them ; but for this wise provision, which keeps the surface dry though surrounded with water, the plants would become overcharged with moisture and rot and decay too rapidly to perfect the ripening of their seeds—a process often carried on at the bottom of streams and lakes, as in the case of the Water-lily and other aquatics. Our blue Iris, however, does not follow this rule, being only partly an aquatic, but stands erect and ripens the large, bony three-sided seeds in a three-sided membraneous pod. The hard seeds of the *Iris versicolor* have been roasted and used as a substitute for coffee. The root, which is creeping, fleshy and tuberous, is possessed of medicinal qualities.

The name Iris, as applied to this genus, was bestowed upon it by the ancient Greeks, ever remarkable for their appreciation of the beautiful, on account of the rainbow tinted hues displayed in the flowers of many of the species ; especially are the prismatic colours shown in the flowers of the large, pearly-white garden Iris, a plant of Eastern origin.

The Fleur-de-lis, as it was formerly written, signified whiteness or purity. This was changed to Fleur-de-luce, a corruption of Fleur-de-Louis—the blossoms of the plant having been selected by Louis the

Seventh of France as his heraldic bearing in the Holy Wars. The flowers of the Iris have ever been favourites with the poet, the architect, and sculptor, as many a fair specimen wrought in stone and marble, or carved in wood, can testify.

The Fleur-de-lis is still the emblem of France.

Longfellow's stanzas to the Iris are very characteristic of that graceful flower :

> " Beautiful lily—dwelling by still river,
>    Or solitary mere,
> Or where the sluggish meadow brook delivers
>    Its waters to the weir.
>
> The wind blows, and uplifts thy drooping banner,
>    And around thee throng and run
> The rushes, the green yeomen of thy manor—
>    The outlaws of the sun.
>
> O fleur-de-luce, bloom on, and let the river
>    Linger to kiss thy feet ;
> O flower of song, bloom on, and make forever
>    The world more fair and sweet."

SHIN-LEAF—SWEET WINTERGREEN—*Pyrola elliptica*—(Nutt.)

> " Wandering far in solitary paths where wild flowers blow,
>   There would I bless His name."—*Heber.*

The familiar name Wintergreen is applied by the Canadians to many species of dwarf evergreen plants, without any reference to their natural affinities. The beautiful family of Pyrolas shares this name in common with many other charming forest flowers on account of their evergreen habit.

Every member of this interesting family is worthy of special notice. Elegant in form and colouring, of a delicate fragrance and enduring verdure, they add to their many attractions the merit of being almost the first green things to refresh the eye, long wearied by gazing on the dazzling white of the snow, for many consecutive months during winter.

As the dissolving crust disappears from the forest, beneath the kindly influence of the transient sunbeams of early Spring, the deep glossy-green shoots of the hardy Pyrolas peep forth, not timidly, as if afraid to meet

> "The snow and blinding sleet ; "

not shrinking from the chilling blast that too often nips the fair promise of April and May; but boldly and cheerfully braving the worst that the capricious season has in store for such early risers.

All bright, and fresh, and glossy, our Wintergreens come forth, as though they had been perfecting their toilet within the sheltering canopy of their snowy chambers, to do honour to the new-born year, just awakening from her icy sleep.

*P. elliptica* forms extensive beds in the forest, the roots creeping with running subterannean shoots, which send up clusters of evergreen leaves, slightly waved and scalloped at the edges, of a deep glossy green and thin in texture.

The name Pyrola is derived from a fancied likeness in the foliage to that of the Pear, but this is not very obvious, nevertheless we will not cavil at it, for it is a pretty sounding word, far better than many a one that has been bestowed upon our showy wild flowers, in compliment to the person who first brought them into notice.

The pale greenish-white flowers of our Pyrola, form a tall terminal raceme ; the five round petals are hollow ; each blossom set on a slender pedicel at the base of which is a small pointed bract ; the anthers are of a reddish orange colour, the stamens ascending in a cluster, while the long style is declined, forming a figure somewhat like the letter J. The seed vessel is ribbed, berry-shaped, slightly flattened and turbinate ; when dry, the light, chaffy seeds escape through valves at the sides. The dry style in this, and most of the genus, remains persistent on the capsule.

The number 5 prevails in this plant ; the calyx is 5 parted ; petals 5 : stamens 10, or twice five ; stigma 1, but 5 rayed with 5 knobs or tubercles at the apex ; seed-vessel 5 celled and 5 valved. The flowers are generally from 5 to 10 on the scape. Most of our Pyrolas are remarkable for the rich fragrance of their flowers, especially *P. elliptica*, and *P. rotundifolia*, together with its variety *incarnata*.

### ONE-SIDED PYROLA—*P. secunda* (L.)

This little evergreen plant is rather singular than pretty. The flowers which are greenish-white form a one-sided slender raceme ; being all turned to one side of the flower-stem ; the style is long and straight, exceeding the stamens and anthers, the latter are very dark, almost dusky black, the stigma, thick and ribbed, forming a turban-shaped green knob in the centre of the flower, stigmas persistent on the capsule. The foliage is dark green, smooth, serrated at the margin of each oval leaf. The leaves are clustered at the base of the flower stem on foot-stalks, leafing the stem upwards a little. The plant is found in dry woods and on banks, under the shade of trees. The flower is scentless.

THE ROUND–LEAVED LESSER PYROLA—*Pyrola rotundifolia* (L.), *v. incarnata* (Gray).

is a far more attractive flower—fragrant with a few sweet pink blossoms and small round or kidney-shaped dark green leaves. Like the sweet Violet of old country hedgerows it betrays its presence by its fine perfume, though often deep hidden among the mosses and weeds which are found in the peat-bogs where it grows. We have yet another Pyrola with round green bell-shaped flowers and dark tipped anthers. This is *Pyrola chlorantha,* (Swartz.)

Though we have none of the Heaths that clothe the hills and common-lands of Scotland and England, we have a large number of beautiful and highly ornamental, as well as useful plants and flowering shrubs belonging to the Natural Order Ericaceæ, which are widely diffused all over the Northern and Eastern portions of the Continent ; wherever there exists a similarity in climate, soil and altitude of the land, there we may expect to find members of the same Natural Orders. Thus we find spread over the Northern and Eastern portions of this Continent, plants that are common to northern European countries ; we have representatives of many familiar flowers, belonging to such families as the Lily, Rose, Violet, Phlox, Saxifrage, Mint, Dogwood, Pyrola, and Campanula, in fact we cannot enumerate the half of what we recognize in our woodlands and plains. It is true that the eye of the botanist will discover some differences in the species, but in most instances these are so little apparent that a casual observer would not notice them. The *Pyrola* has its representative flower in England. The *Linnæa,* in Norway. Our pretty *Smilacina bifolia,* or " Wild Lily of the Valley," and our Low Cornel are also found with many of our native Ferns, in that Northern land of mountain, flood and forest.

It is pleasant to recognize an old familiar flower, it is like the face of an old friend in a foreign country, bringing back the memory of days lang syne, when the flowers that we gathered in our childhood were a joy and a delight to heart and eye.

ONE-FLOWERED PYROLA—*Moneses uniflora* (Gray).

This exquisitely scented flower is only found in the shade of the forest, in rich, black, leaf mould, where, like *P. elliptica,* it forms considerable beds ; it is of evergreen habit. The leaves are of a dark green and smooth surface, clustered at the base of short stems which rise from the running root-stock, from the centre of each of which rises one simple scape, bearing a gracefully nodding flower ; each milk-white petal is elegantly scalloped ; the stamens, eight to ten, are set close to the base of the petals ; the anthers are of a bright purple-amethyst colour ; the style straight, with five radiating points at the extremity, forming a perfect mural crown in shape ; it is bright green, and much exceeds in length the stamens.

The scent of the flower is very fine, resembling in richness that of the Hyacinth.

The members of the Pyrola family are, for the most part, found in rich woods, some in low, wet ground, but a few prefer the drier soil of fôrests ; one of these is the exquisitely beautiful evergreen plant known by Canadian settlers as

PRINCE'S PINE—*Chimaphila umbellata* (Nutt),

From root to summit this plant is altogether lovely. The leaves are dark, shining and smooth, evergreen and finely scrrated ; the stem is of a bright rosy-red ; the delicately pink-tinted flowers look as if moulded from wax : the anthers are of a bright amethyst-purple, set round the emerald-green turbinated stigma. The flowers are not many, but form a loose corymb springing from the centre of the shining green leaves. There is scarcely a more attractive native plant than the *Chimaphila* in our Canadian flora.

The leaves of this beautiful Wintergreen are held in high estimation by the Indian herbalists who call it Rheumatism weed *( Pipsissewa )*. It is bitter and aromatic in quality.

LUPINE—*Lupinus perennis* (L.)
(PLATE IV.)

" Lupine whose azure eye sparkles with dew."

Those who know the Blue Lupine only as a cultivated flower can form but a poor idea of its beauty in its wild state on the rolling prairies or plainlands.

On light loamy or sandy soil our gay Lupine may be seen, gladdening the wastes and purpling the ground with its long spikes of azure blue, white and purple flowers, of many shades.

The Lupine comes in with the larger yellow Moccasin (*Cypripedium pubescens*); the *Trillium grandiflorum*; the white Pyrola, Wild Rose (*Rosa blanda*); Scarlet-cup (*Castilleia coccinea*) and many others in the flowery month of June ; mingling its azure flowers with these, it produces an effect most pleasing to the eye.

The blossoms, like those of all the Pulse tribe to which it belongs, are papilionaceous or winged. The two upper petals or wings are concave, closing over the scythe shaped keel, which encloses the stamens, these are united into a bundle at the base (this arrangement is called by botanists monadelphous). The sheath that conceals the stamens is entire, pointed and varying in colour from white to reddish-purple. The flowers are set on short pedicels or flower stalks, forming a close, long, terminal raceme, the lower flowers opening first. The stem is leafy, erect, downy; the leaves on longish foot stalks are composed of from seven to nine soft, greyish, silky leaflets, set round the central axis of the stalk in a horizontal circle. The whole plant is soft and velvety in appearance. The pods are long and somewhat broad. The seeds are ivory white when fully ripe, and are the food of Squirrels, Partridges, Field-mice and other wild denizens of the wilderness. The Lupine can be readily grown from seed, and blooms well in our garden plots, abiding with us year after year. The ivory white seeds are often introduced into those pretty fanciful wreaths, frequently exhibited at our township shows, and known as the "Farmer's Wreath," being composed of different varieties of grain and seeds, arranged so as to form flowers, leaves, fruits, &c.

Before the plainlands above Rice Lake were enclosed and cultivated, the extensive grassy flats were brilliant with the azure hues of the Lupine in the months of June and July ; but the progress of civilization sweeps these fair ornaments from the soil. What the lover of the country loses of the beautiful, is gained by the farmer in the increase of the useful, and so it must be ; but nevertheless we mourn for the beautiful things which gladdened our eyes.

"Oh wail for the forest its glories are o'er."

TWIN-FLOWER—*Linnæa borealis* (Gronov.)

" Nestled at its roots is beauty,
    Such as blooms not, in the glare
    Of the broad sun. That delicate forest flower
    With scented breath, and look so like a smile,
    Seems, as it issues from the shapeless mould,
    An emanation from the indwelling life."—*Bryant*.

" And there Linnæa weaves her rosy wreath."

This delicate and graceful little evergreen is widely diffused through most of the Northern countries of Europe and America. It is found within the limits of the Arctic Circle : in dreary Kamschatka, and in snowy Lapland, the young girls wreathe their hair with its flexible garlands. In inhospitable Labrador it covers the rocks and mossy roots of Pines and Birches in lonely shaded glens. It is found in the Scottish Highlands and through all parts of the Northern and Eastern States of America. In all the Provinces of our own Canada it may be found in secluded spots. On the rocky Islands of the St. Lawrence, and of our inland lakes it is particularly abundant, and its graceful trailing branches cover

D

the rude rocks, and fling a robe of luxuriant vegetation over decaying
fallen timber, concealing that which is unseemly with grace and beauty.

> " Sweet flower, that in the lonely wood
>    And tangled forest, clothest the rude twisted roots
>    Of lofty pine and feathery hemlock,
>    With thy flower-decked garland ever green ;
>    Thy modest, drooping, rosy bells of fairy lightness
>    Wave gently to the passing breeze,
>                    Diffusing fragrance."

This pretty, graceful little plant was named in honour of the great
father of botany, the good Linnæus, who chose it more especially as
his own flower when he plucked it first in Bothnia : by his wish it was
adopted for the crest of his coat of arms.

The little flower has been immortalised by the great botanist. It
is said that one of his pupils aware of his great master's love for the
plant, when visiting China, caused a service of fine porcelain to be made
and decorated with wreaths of the Linnæa, as a present to Linnæus, and
as a mark of his grateful remembrance.

At the death of the great naturalist, Cardinal de Noailles erected a
cenotaph in his garden, to his memory, and planted this little northern
flower at its base for the sake of him whose name it bears.*

At every joint the Linnæa puts forth white, fibrous rootlets, thus increasing and perpetuating the growth of the plant till it forms a tangled,
mass of leafy branches. The leaves are round, slightly crenate with
a deeper notch at the top, and together with the younger stalks are somewhat hairy. They are placed in opposite pairs, from the centre of each
of which rises a slender flower stalk, forking near the summit, and bearing
a pair of delicate, rose-tinted drooping bells, veined with lines of a
deeper pink. The throat of the bell is tubular, as in the Honey-suckle
and is thickly beset with silvery, woolly hairs. Stamens four, two of
them shorter than the others ; the corolla is divided near the margin into
five pointed segments. Seed vessel a dry, three-celled, but one-seeded
pod.

If planted for cultivation, the ground should be shaded and somewhat damp. In an artificial rock-work, sufficiently protected from the
glare of sunshine and kept moist in hot days, it would grow luxuriantly
and throw its evergreen matted branches over and among the stones with
pretty effect. The blossoms give out a delicate fragrance, especially at
dewfall, the scent being scarcely perceptible during the noontide heat.

---

* See Miss Brightwell's Life of Linnæus,

Our charming Twin-flower is very constant in its habits, being found year after year in the same locality, as long as it enjoys the advantages of shade and moisture ; but it cannot endure exposure to the heat and glare of sunshine, though it will linger as long as it can obtain any shelter.

Thirty years ago I found the *Linnæa borealis* growing beneath the shade of Hemlock trees among long Sphagnous mosses, on the rocky banks of the Otonabee. Last year, on re-visiting the same spot, I noticed a few dwarfed, yellow and starved-looking plants struggling, as it were, for existence, but the evergreens that had sheltered them at their roots were all gone.

There seems to be a law of mutual dependence among the vegetable tribes, each one ministering to the wants of the others. Thus the shelter afforded by the larger trees to the smaller shrubs and herbs, is repaid again to them by the nourishment that the decaying leaves and stems of these latter afford, and the warmth that they yield to their roots by covering the ground from the winter cold, and thus protecting them from injury. Further than this, it is very probable that they appropriate to their own use qualities, in the soil or in the air, that might prove injurious to the healthy growth of the larger vegetables. That which is taken up by one race of plants is often rejected by others. Yet so beautiful is the arrangement of God's economy in the vegetable world that something gathers up all fragments and nothing is lost—nay, not the minutest particle runs to waste. The farmer practically acknowledges the principle that one kind of vegetable feeds upon that which another rejects, when he adopts a certain routine in cropping his land, for he knows that if he planted grain in constant succession the soil would soon cease to yield its increase, because it would have ceased to afford the food necessary for perfecting the grain : but he sows Wheat after roots, as Potatoes, Turnips and Beets, or after pulse as Pease, Beans or Vetches, for these have taken only certain constituents of the soil, leaving those portions on which the Cereals feed unappropriated. Thus silently, unconsciously, and mysteriously, do God's creatures administer to one another, working out the will of their Great Creator, and obeying his laws while following the instincts of their several natures.

We might follow out this subject to a greater length than our limits will admit of our doing, but it is time that we dismiss our lovely little Twin-flower which forms so attractive a feature in our artist's graceful design, hoping that it may sometimes win an admiring glance from our readers, who may be so fortunate as to meet with its evergreen wreaths and fragrant flowers, in its native woods during the leafy month of

June, which is its flowering season—though often it may be seen
lingering in rocky woods through July, and now and then a few late
blossoms will be found in shady ground late in August.

### ROUND-LEAVED SUNDEW—*Drosera rotundifolia* (L.)

Two species of this interesting and singular family are common in
Canada.   One *Drosera rotundifolia* with round leaves, beset with stiff
glandular hairs of a deep red colour, abounds in boggy soil in most
parts of the Dominion.

The beauty of this little plant consists in the hairy fringes of the
leaves which exude drops of a clear dew-like fluid ; each little leaf seems
adorned with a row of liquid gems, beautiful as pearls, and glistening in
the sunlight like miniature diamonds.

The round red leaves are prolonged into the petiole, or rather the
leaf-stalk is expanded at its edges and terminates in the glandular leaf.
The flowers are small, white, sometimes tinged with pink, borne on a
slender, naked, somewhat one-sided scape, which droops a little at the
tip.   I am not aware of any medicinal or useful qualities of the
Sundews, but the eye that sees the beauty set forth in the little dew-
gemmed leaf of this lovely plant, may behold in it with reverent
admiration a work of creative mind, surpassing all that man's ingenuity
can produce.   The jeweller may polish and set the ruby and the
diamond in fretted gold, but he cannot make one ruby-tinted leaf of
the little Sundew.

A rather narrower-leaved species is *Drosera longifolia* (L.), which
grows abundantly in a peat marsh near Stoney Lake, at a spot known as
" Hurricane Point," a rocky cape, at the rear of which lies a low marshy
flat, covering several acres of wet ground ;  a rare garden and nursery
for many charming flowering shrubs and exquisite bog-loving plants.
A beautiful carpet of white Peat Moss *Sphagnum cymbifolium* is spread
over the surface, nearly a foot deep ; on this we see wreaths of the graceful low-bush Cranberry, trailing its slender branches with their dark green
glossy myrtle-like foliage and delicate pink revolute flowers, as well as
berries in every stage of progress, the tiny green immature fruit—the
golden—the mottled and the deep red ripe berry.   How tempting to the
hand and eye.   There the slender leaved Sundew mixes its white
flowers with the fringed Orchis, and sends up from the watery soil its
modest flowers in the midst of a bed of the grand blossoms of that
rarely constructed plant the " Pitcher Plant," *Sarracenia purpurea*, or as
it is called by some writers " Side-saddle flower."*

---

* Gray says it is difficult to fancy any resemblance between this flower and a side-saddle.  I
venture to suggest that the common name originated from the flap-like extension of the leaf.

The bog of which I speak abounds in shrubs, among which we see the narrow, dark-leaved Sheep-laurel, *Kalmia glauca*, with its rose-coloured flowers ; the aromatic Sweet-Gale, *Myrica Gale ;* and Labrador-tea, *Ledum latifolium* with its revolute, rosemary-like, narrow leaves, and whitish flowers.    Above all, for beauty, is the White Peat Moss itself, with its soft, velvety foliage, varying in shade from pale sea-green or creamy-white to delicate pink and deeper rose.   I know of nothing more lovely than are these exquisite *Sphagnums .* nor are they without their value, for they are greatly used by the florist and gardener in packing roots and plants for sale.

There are more vegetable treasures to be found in the peat marsh near Hurricane Point than I have noticed.   A deer track leads beyond this marsh to " Fairy Lake."   This lake is like a mountain tarn ; it is surrounded by lofty rocks, and is not a mere inlet from Stoney Lake, as it now appears, being encircled on all sides by a stony barrier of rugged rocks, some rising from the water's edge, bare and precipitous, or clothed with grey, hoary tufts of *Cladonias* and other lichens and mosses.   In the clefts may be found the somewhat rare *Woodsia Ilvensis*,  Hairy Woodsia,  and  the  Rock  Polypody, *P. vulgare.*   The last named is not, indeed, an uncommon adornment to the rocky bluffs and stony islands of our back lakes, and enlivens the rugged, grey, rocky surfaces with its bright, glossy fronds and golden fruit dots.   The rocks decline to the side facing the larger lake, and towards the western corner there is a bed of the White Peat Moss, overshadowed by a forest of that grand fern, *Osmunda regalis*, worthy of its regal name, for here, among the soft Sphagnum, and towering to the height of five and six feet, it bears above its light green leafage (or should I say *frondage ?*) its rich tufts of cinnamon-brown sporangia.   Beneath the Osmundas, and rising above the mosses, the crimson-lipped leaves and large, red flowers of the "Pitcher Plant" *Sarracenia purpurea*, may be seen in great perfection.

These are but a few of the attractions of Fairy Lake, for there are flowers and flowering shrubs, that grow in the wild, rocky soil, of many kinds.   The beautiful spikes of the rose-blossomed *Spirea tomentosa*, the Hardhack of the Indians, and the graceful white *Spirea salicifolia*, wild Roses, and Golden-rods, and Asters, with many others are scattered round this lovely lakelet, rendering it a place of interest to the botanist and to the pleasure-seeking tourist.

PITCHER PLANT—SOLDIER'S DRINKING CUP—*Sarracenia purpurea* (L.)

Even the most casual observer, in passing a bed of these most remarkable plants, must be struck by their appearance.   Indeed, from root to flower, they are in every way worthy of our notice and admiration.

The Pitcher Plant is by no means one of those flowers found only in inaccessible bogs and dense cedar-swamps, as are some of our rare and lovely Orchids. In almost any grassy swamp, at the borders of low-lying lakes, and beaver-meadows—often in wet, spongy meadows—it may be found forming large beds of luxuriant growth.

When wet with recent showers, or glistening with dew-drops, the rich crimson veinings of the broadly scalloped lip of the tubular leaf, (which is thickly beset with fine stiff silvery hairs,) retain the moisture, and shine and glisten in the sun-light.

The root-stock is thick, and bears many fibres. The tubular leaves are of a reddish tinge on the outer and convex side, but of a delicate light-green within. The texture is soft, smooth and leathery; the base of the leaf, at the root, is narrow and pipe-stem-like, expanding into a large hollow receptacle, capable of containing a wine-glass-full of liquid; even in dry seasons this cup is rarely found empty. The hollow form of the leaves, and the broad ewer-like lips, have obtained for the plant its local and wide-spread name of "Pitcher Plant," and "Soldier's Drinking Cup." The last name I had from a poor old emigrant pensioner, when he brought me a specimen of the plant from the banks of a half dried up lake, near which he was located : " Many a draft of blessed water have we poor soldiers had, when in Egypt, out of the leaves of a plant like this, and we used to call it the " Soldier's Drinking Cup."

Most probably the plant that afforded the "blessed water" to the poor thirsty soldiers was, the *Nepenthes distillatoria*, which plant is found in Egypt and other parts of Africa. Perhaps there are but few among the inhabitants of this well-watered country that have as fully appreciated the value of the Pitcher Plant as did our poor uneducated Irish pensioner, who said that he always thought that God in His goodness had created the plant to give drink to such as were athirst on a hot and toilsome march ; and so he looked with gratitude and admiration on its representative in Canada. Many a lesson may we learn from the lips of the poor and the lowly.

Along the inner portion of the leaf there is a wing or flap which adds to its curious appearance. The evident use of this appendage is to contract the inner side of the leaf, and to produce a corresponding rounding of the outer portion, which is thus thrown backwards, and enables the moisture more readily to fill the cup and to be there retained. Quantities of small flies, beetles and other insects, enter the pitcher, possibly for shelter, but are unable to get out again, owing to the reflexed bristly hairs that line the upper part of the tube and lip, and thus find a watery grave in the moisture that fills the hollow below, whence there is no escape for the poor deluded prisoners.

The tall stately blossom of the Pitcher Plant is not less worthy of our attention than the curiously formed leaves. The smooth, round, simple scape rises from the centre of the plant to the height of eighteen inches or two feet. The flower is single and terminal, composed of five sepals, with three little bracts ; five blunt, broad petals of a dull purplish red colour, but sometimes red and light-yellowish green ; and in one variety the petals are mostly of a pale-green hue, and there is an absence of the crimson veins in the leafage. The petals are incurved or bent downwards towards the centre. The stamens are numerous. The ovary is five-celled, and the style is expanded at the summit into a five-angled, five-rayed, umbrella-like scalloped mantle, which conceals beneath it five delicate rays, each terminating in a little hooked stigma. The capsule or seed-vessel is five celled and five-valved ; seeds numerous.

I have been more minute in the description of this interesting plant, because much of its peculiar organization is hidden from the eye, and cannot even be recognized in a drawing, unless it be a strictly botanical one, with all its interior parts dissected ; and also because the Pitcher Plant has lately attracted much attention by its reputed medicinal qualities in cases of Small-pox, that loathsome scourge of the human race. A decoction from the root of this plant has been said to lessen all the more violent symptoms of the disorder. If this be really so, its use and application should be widely known ; fortunately the remedy would be within the reach of everyone; like many of our sanative herbs it is to be found without difficulty, and being so remarkable in its appearance, can never be mistaken by the most ignorant of our country herbalists for any injurious substitute.*

### WILD ORANGE LILY—*Lilium Philadelphicum* (Lin.)

" Consider the lilies of the field, how they grow ; they toil not, neither do they spin ; and yet I say unto you, that Solomon in all his glory was not arrayed like one of these."

The word Lily is said to be derived from a Celtic word, *Li*, which signifies whiteness; also from the Greek, *Lirion*. Probably the stately Lily of the garden, *Lilium candidum*, was the flower to which the name was first given, from its ivory whiteness and the exquisite polish of its petals. However that may be, the name Lily is ever associated in our minds with grace and purity, and reminds us of the Saviour of men, who spake of the Lilies of the field, how they grew and flourished beneath the care of Him who clothed them in robes of beauty more gorgeous than the kingly garments of Royal Solomon

---

* NOTE.—I regret to be compelled to say that later experience has dispelled belief in the virtue of the Pitcher Plant, no such good results having been obtained from repeated trials in cases of that direful disease, small-pox.

Sir James Smith, one of the most celebrated of English botanists, suggests that the Lilies alluded to by our Lord may have been *Amaryllis lutea*, or the Golden Lily of Palestine—the bright yellow blossoms of this plant abound in the fields of Judæa, and at that moment probably caught His eye, their glowing colour aptly illustrating the subject on which He was about to speak.

The Lily has a wide geographical range, and may be found in some form in every clime.

There are Lilies that bloom within the cold influence of the frigid zone, as well as the more brilliant species that glow beneath the blazing suns of the equator in Africa and Southern Asia.

Dr. Richardson mentions, in his list of Arctic plants, *Lilium Philadelphicum*, our own gorgeous orange (or rather scarlet spotted) Lily. He remarks that it is called by the Esquimaux " Mouse-root." from the fact that it is much sought after by the field mice, which feed upon the root. The Porcupine also digs for it in the sandy soil in which it delights to grow.

In Kamtschatka the *Lilium pomponium* is used by the natives as an article of food ; and in Muscovy the white Narcissus is roasted as a substitute for bread.

The healing qualities of the large white Lily roots and leaves, when applied in the form of a poultice to sores and boils, are well known. Thus are beauty and usefulness united in this most attractive plant.

We find the Orange Lily most frequently growing on open plain-lands, where the soil is sandy loam. In partially-shaded grassy thickets in oak-openings, in the months of June and July, it may be seen mixed with the azure blue Lupine (*Lupinus perennis*), the golden-flowered Moccasin (*Cypripedium pubescens*), the large sweet-scented Wintergreen (*Pyrola elliptica*), and other charming summer flowers. Among these our gay and gorgeous Lily stands conspicuous.

The stem is from eighteen inches to two feet high. The leaves are narrow pointed, and of a dark green colour, growing in whorls at intervals round the stem. The flowers are from one to three ; large open bells, of a rich orange scarlet, within, spotted with purplish brown or black. The outer surface of the petals is pale orange ; anthers six, on long filaments ; pollen of a brick red, or brown colour ; stigma three-lobed.

Many flowers increase in beauty of colour and size under cultivation in our gardens, but our glorious Lily can hardly be seen to greater advantage than when growing wild on the open plains and prairies, under the bright skies of its native wilderness.

HAREBELL—*Campanula rotundifolia* (Lin.)

" With drooping bells of purest blue
Thou didst attract my childish view,
    Almost resembling
The azure butterflies that flew,
Where 'mid the heath thy blossoms grew,
    So lightly trembling."

The writer of the above charming lines has also called the Harebell " the Flower of Memory," and truly the sight of these fair flowers, when found in lonely spots in Canada, has carried one back in thought to the wild heathery moors or sylvan lanes of the mother country.

" I think upon the heathery hills
    I ae hae lo'ed sae dearly ;
I think upon the wimpling burn
    That wandered by sae clearly."

But sylvan wooded lanes, and heathery moorlands are not characters of our Canadian scenery, and if we would find the Harebell, we must look for it on the dry gravelly banks of lakes and rivers, or on rocky islets, for these are its haunts in Canada.

Although, in colour and shape of the blossom, the Canadian flower resembles the British one, and is considered by botanists to be the same species ; it is less fragile, the flower stems being stouter, and the foot stalk or pedicel stiffer and less pendulous ; the root leaves, which are not very conspicuous during its flowering season, are round, heart-shaped. Those of the flower-stem are numerous, narrow, and pointed. This pretty flower is variable in colour and foliage. Its general flowering season is July and August.

The corolla is bell-shaped or campanulate, five cleft ; calyx lobes, awl shaped, persistent on the seed vessel ; stamens five ; style one ; stigmas two ; seed vessel several celled and many seeded ; in height the plant varies from a few inches to a foot ; number of flowers varying from a few to many.

We have three common species in Canada, the present one ; *Campanula Americana,* (Lin.) a large, handsome species, found in Western Canada ; and *C. aparinoides*, (Pursh) the Rough-leaved Bellflower which is found in thickets and swamps, it is of a climbing or rather clinging habit, the weak slender stem, many branched, laying hold of the grasses and low shrubs that surround it, for support, which its rough teeth enable it to do very effectually : in habit it resembles the smaller *Galium*, or Lady's Bed-straw. The graceful bell-shaped flowers are of a delicate lavender colour. The leaves of this species are narrow-linear, rough with minutely-toothed bristles ; the flowers are few, and fade very quickly. The name Campanula is from *campana,* a bell.

The Harebell has often formed the theme of our modern poets, as illustrative of grace and lightness. In the Lady of the Lake we have this pretty couplet when describing Ellen :

> " E'en the light Harebell raised its head
> Elastic from her airy tread."

### YELLOW-FLOWERED WOOD-SORREL—*Oxalis stricta* (L.)

This delicate little flower may be found occasionally by the wayside; but is oftener seen among the herbage near the borders of cultivated fields. The trifoliate leaves are terminal on longish foot-stalks, thin in texture, and of a pleasant acid taste. At sunset, like the clover and other trefoils, it droops and folds its leaflets together to sleep, for some plants rest as in sleep. This Wood-sorrel is somewhat branching and bushy ; the pale yellow blossoms are on long stalks, fading very soon. There is also another species—*Oxalis Acetosella* (L.)—white with purple veinings, a lovely delicate thing of great beauty, which is found on damp mossy banks at the edge of low pastures. It has been asserted by some persons that the Wood-Sorrel is the Irish Shamrock, the emblem of the Holy Trinity ; but it is more likely, if St. Patrick really used any plant as a simile, that he took the familiar golden-blossomed trefoil Yellow-clover, which is " The Shamrock," which grows so abundantly in Ireland by waysides. The Wood-sorrel is of rarer occurrence and of less familiar appearance.

### CISTUS—ROCK-ROSE—*Helianthemum Canadense* (Michx.)

We find the yellow Cistus growing on gravelly hills and sunny banks. It is a pretty, delicate-flowered plant of slender upright growth, and hoary foliage, beset with silvery gray hairs. The flowers, rarely more than two opening at a time, are about an inch wide ; the petals slightly notched at the upper edge, of a pale brimstone colour ; the many stamens and anthers reddish-orange. The flowers open at sunrise but fall before night ; they are so slight in texture that the least touch bothers them. There is a peculiarity in this plant that is very singular, the tendency to produce an abundance of abortive flowers along the lower portion of the stem. These never open, and give a scaly look to the plant. The Cistus is also known by the name of Frost-Plant ; this name may have been given to it from the hoary appearance of the leaves, though a less obvious cause has been assigned for the name. It is said that ice crystals are formed on the bark in the autumnal frosts ; but most likely some crystallized substance from the juices of the plant has been mistaken for ice.

YELLOW-FLAX—WILD FLAX—*Linum sulcatum* (Riddell.)

This is a delicate little plant mostly found on dry sunny banks, during the hot summer months. The blossoms resemble the common blue Flax, but are smaller; the narrow leaves are harder in texture and the plant not more than one foot in height; the flower falls very soon. I do not know if the stem possesses the thready flax fibre of the cultivated species; its only recommendation is the pretty pale yellow blossom.

CANADIAN BALSAM—*Impatiens fulva* (Nutt.)

Our Wild Balsam is a singularly gay plant, with its profusion of orange-coloured spotted flowers, light foliage and semi-transparent stems. The Butterflies seem to take delight in hovering over the bright blossoms, and the Humming-birds may be seen on sunny days with outstretched beaks and wings, winnowing the air as they balance their tiny bodies, while extracting sweets or insects from the curiously hooded flowers. In the New England States it is known as the Humming-bird Flower, but it has other pretty descriptive names, Jewel Weed, Speckled Jewel, and Touch-me-not. This last alludes to the sensitive nature of the slender seed-pods, which burst at a slight touch, rolling themselves into pretty rings and shedding abroad the seed.

The flowers hang lightly, drooping on very slender thready stalks; when open the outer sepal of the coloured calyx forms a hooded cap which reminds one of an old jester's cap and bells. It is only in the single-flowered Balsam, under cultivation, that we see the curious hood with its horn-like nectary; but the elastic seed-pod is, like the wild species, equally sensitive if touched. A strong colouring matter of bright orange pervades the whole plant in our Wild Balsam—leaves, stem and flower. The Indian women use the juice in dyeing, and also apply it in Erysipelas caused by Poison Ivy and in other diseased states of the skin. Our Balsam loves low wet soil. The low lake shore, and forest streams are its favourite haunts, where it attains the height of three and four feet.

There are two species, *Impatiens fulva*, distinguished by its deeper coloured blossoms, orange, almost scarlet, and its brown spots, and darker green leaves. *I. pallida* (Nutt) is paler, and the markings on the petals slighter, the foliage is much lighter, and the juice of the plant more watery.

Professor Lindley has given the Balsam a place among the garden Nasturtiums. A very natural affinity seems to exist between the Nasturtiums and Balsams as respects habits, form and colour. Dr. Gray gives the Balsams an order to themselves.

RATTLESNAKE PLANTAIN—*Goodyera pubescens* (R. Br.)

This is a formidable name for a lovely little plant, the leaves of which are prettily netted over the dark green surface with milky-white veinings.   The ovate, pointed leaves are set close to the ground; from the centre of the leaves rises a naked stalk of pearly white flowers in a slender spike; corolla ringent with inflated lip; root-stock somewhat creeping, soft and fibrous; the flowers are slightly fragrant.   This pretty little plant is found in the forest, often on fallen decayed trunks of trees, or in light fibrous mould.   It is very nearly allied to the

SLENDER LADIES'-TRESSES—*Spiranthes gracilis* (Big.)

The flower-stem of this singular plant is twisted, so that the blossoms are turned to one side, forming a spiral of great beauty.   The flowers are larger than those of the Rattlesnake Plantain, and sweeter; greenish-white, lipped and fringed.   The two leaves are closely pressed to the ground, and are little seen after the plant is in bloom.   There are several species of these graceful Orchids.

The spiral arrangement of the flowers probably suggested the ringlets on some fair lady's head.   The old florists and herbalists of former times were more gallant than our modern botanists, for they gave many pretty names to the flowers instead of the harsh-sounding, unmeaning ones that we find in our scientific manuals of Botany.   So we have among our local and familiar names, such prettily sounding ones, as "Ladies' Tresses," "Sweet Ciccly," "Sweet Marjoram," or "Marjory," "Mary-gold," "Ladies' Slipper," with a number of others that I could name—besides descriptive names, which form a sort of biography of the plant, giving us a correct idea of their characteristics and peculiar uses or habits.

SWEET SCENTED WATER-LILY—*Nymphæa odorata* (Ait.)

" Rocked gently there, the beautiful Nymphæa
Pillows her bright head."—*Calendar of Flowers.*

Water-Lily is the popular name by which this beautiful aquatic plant is known, nor can we find it in our hearts to reject the name of Lily for this ornament of our lakes.   The White Nymphæa might indeed be termed "Queen of the Lakes," for truly she sits in regal pride upon her watery throne, a very queen among flowers.

Very lovely are the Water Lilies of England; but their fair sisters of the New World excel them in size and fragrance.

Many of the tribe, to which these plants belong, are natives of the torrid zone, but our White Water-Lilies (*Nymphæa odorata* and *tuberosa*) and the Yellow Pond-Lilies (*Nuphar advena, lutea* and *Kalmiana*) only are able to support the cold winters of Canada. The depth of the water in which they grow enables them to withstand the cold, the frost rarely penetrating to their roots, which in the Nymphæas are rough and knotted, white and fleshy, and often as thick as a man's wrist. The root-stock is horizontal, sending many fibrous slender rootlets into the soft mud; the stems that support the leaves and blossoms are round, of an olive-green, containing open pores filled with air, which cause them to be buoyed up in the water. These air-cells may be examined by cutting the stems across, when the beautiful arrangement of the pores can be seen, and admired for the use they are, in buoying up the stem, and allowing the flower-cup to float upon the surface of the water. These air-cells are arranged with beautiful symmetry, so as to give strength as well as lightness.

The leaves of the Water-Lily are of a full-green colour, deeply tinged with red towards the fall of the year, so as to give a blood red tinge to the water ; they are of a large size, round kidney-shaped, of leathery texture, and highly-polished surface ; resisting the action of the water as if coated with oil or varnish. Over these beds of Water-Lilies hundreds of Dragon Flies of every colour—blue, green, scarlet and bronze—may be seen like living gems, flirting their pearly-tinted wings in all the enjoyment of their newly found existence ; possibly enjoying the delicious aroma from the odourous lemon-scented flowers, over which they sport so gaily.

The flowers of the Water-Lily grow singly at the summit of the round, smooth, fleshy scapes. Who that has ever floated upon one of our calm inland lakes, on a warm July or August day, but has been tempted, at the risk of upsetting the frail birch-bark canoe, or shallow skiff, to put forth a hand to snatch one of those matchless ivory cups, that rest in spotless purity upon the tranquil water, just rising and falling with the movement of the stream ; or has gazed with wishful and admiring eyes into the still, clear water, at the exquisite buds and half unfolded blossoms that are springing upwards to the air and sun-light.

The hollow boat-shaped sepals of the calyx are four in number, of a bright olive green, smooth and oily in texture. The flowers do not expand fully until they reach the surface. The petals are numerous, hollow (or concave), blunt, of a pure ivory white ; very fragrant, having the rich odour of freshly-cut lemons ; they are set round the surface of the ovary (or seed-vessel) in regular rows, one above the other, gradually lessening in size, till they change, by imperceptible gradation, into the narrow, fleshy petal-like yellow anthers. The pistil is without style, the stigma forming a flat rayed top to the ovary, as in the Poppy and many other plants.

But if the White Water-Lily is beautiful, how much more so is the lovely pink-flowered variety shown in our Plate, which was painted by Mrs. Chamberlin from a specimen she collected at Lakefield and which was of such an exquisite shade of colour that it could be only compared with the

> " Hues of the rich unfolding morn,
> That ere the glorious sun be born
> By some soft touch invisible,
> Around his path are taught to swell."—*Keble.*

This is called *N. odorata* var. *rosea* and is found abundantly in many of the small lakes in the northern counties of Ontario, particularly in the Muskoka district.

On the approach of night our lovely water-nymph gradually closes her petals, and slowly retires to rest on her watery bed, to rise again the following day, to court the warmth and light so necessary for the perfection of the embryo seeds, and this continues till the fertilization of the germ has been completed, when the petals shrink and wither, and the seed-vessel sinks down to the bottom of the water, where the seeds ripen in its secret chambers. Thus silently and mysteriously does Nature perform her wonderful work, " sought out only by those who have pleasure therein."*

The roots of the Water-Lily contain a large quantity of fecula (flour), which, after repeated washings, may be used for food; they are also made use of in medicine, being cooling and softening; the fresh leaves are used as good dressing for blisters.

The Lotus of Egypt belongs to this family, and not only furnished magnificient ornaments with which to crown the heads of their gods and kings, but the seeds also serve as food to the people in times of scarcity. The Sacred Lotus (*Nelumbium speciosum*) was an object itself of religious veneration to the ancient Egyptians.

The Chinese, in some places of that over-populated country, grow Water-Lilies upon their lakes for the sake of the nourishment yielded by the roots and seeds.

" Lotus-eaters," says Dr. Lee, " not only abound in Egypt, but all over the East." " The large fleshy roots of the *Nelumbium luteum*,

---

* In that singular plant, the Eel or Tapegrass *Vallisneria spiralis* (L.) a plant indigenous to our slow-flowing waters, the elastic stem which bears the pistillate flowers uncoils to reach the surface of the water: about the same time the pollen-bearing flowers, which are produced at the bottom of the water on very short scapes, break away from the confining bonds that hold them, and rise to the surface, where they expand and scatter their fertilizing dust upon the fruit-bearing flowers which float around them; these, after a while, coil up again and draw the pod-like ovary down to the bottom, there to ripen and perfect the fruit.

or great Yellow Water-Lily, found in our North American lakes resemble the Sweet Potato (*Batatas edulis*), and by some of the natives are esteemed equally agreeable and wholesome," observes the same author, "being used as food by the Indians, as well as some of the Tartar tribes."

As yet little value has been attached to our charming White Water-Lily, because its uses have been unknown. It is one of the privileges of the botanist, and naturalist, to lay open the vegetable treasures that are so lavishly bestowed upon us by the bountiful hand of the great Creator.

YELLOW POND-LILY—SPATTER DOCK—*Nuphar advena* (Ait.)

> And there the bright Nymphæa loves to lave,
> And spreads her golden orbs along the dimpling wave.

The Yellow Pond-Lily is often found growing in extensive beds, mingled with the White, and though it is less graceful in form, there is yet much to admire in its rich orange-coloured flowers, which appear, at a little distance, like balls of gold floating on the still waters. The large hollow petal-like sepals that surround the flower, are sometimes finely clouded with dark red on the outer side, but of a deep orange yellow within, as also are the strap-like petals and stamens : the stigma, or summit of the pistil, is flat, and 12-24 rayed. The leaves are dark-green, scarcely so large as those of the White Lily, and more elongated, they are borne on long thick fleshy stalks, flattened on the inner side, and rounded without. The botanical name *Nuphar* is derived, says Gray from the Arabic word *Neufar*, signifying Pond-Lily.

Nature's arrangements are always graceful and harmonious, and this is illustrated by the grouping of these beautiful water plants together. The ivory white of the large Lily mingling with the brighter, more gorgeous colour of the yellow ; and the deeper green of the broad shield-like leaf with the bright verdure of that of the Arrow-head, and the bright rosy tufts of the red Water Persicaria ; the leaves, veinings and stems, giving warm tints of colour to the water, as they rise and sink with the passing breeze.

Where there is a deep deposit of mud in the shallows of still waters, we frequently find many different species of aquatics growing promiscuously. The tall lance-like leaf and blue-spiked heads of the stately *Pontederia cordata*, keeping guard, as it were, over the graceful Nymphæa, like a gallant knight with lance in rest, ready to defend his queen; and around these the fair and delicate white flowers of the small Arrow-head rest their frail petals upon the water, looking as if the slightest breeze that ruffled its surface, would send them from their watery pillow.

Beyond this aquatic garden lie beds of Wild Rice (*Zizania aquatica*) with floating leaves of emerald green, and waving grassy flowers of straw-colour and purple—while nearer to the shore the bright rosy tufts of the Water Persicaria (*Polygonum amphibium*), with dark-green leaves and crimson stalks, delight the eyes of the passer-by.

### SPIKENARD—*Aralia racemosa* (L.)

This valuable plant is distinguished by its heart-shaped, five-foliate, pointed and serrated leaves ; wide-branching, herbaceous stem ; long, white, aromatic, astringent root ; greenish white flowers and racemose branching umbels of small, round, purple berries, about the size and colour of the purple-berried elder.   It affects a rich, deep soil, the long, tough roots sometimes extending to a yard or more in length ; forking and branching repeatedly.   The plants are often seen growing on large boulders where there is a sufficiency of soil, the roots penetrating into the crevices, or extending horizontally over the surface.   Another favourite place for this plant is on the earth adhering to large upturned roots, the seed having been left by the birds.   The root has an aromatic taste, and smells like Aniseed or Caraway.  It is a most valuable domestic medicine, safe and simple ; its curative properties, in cases of obstinate dysenterical disorders, deserve to be widely known.

It was from an old Canadian settler that I learned the virtue of the *Spignet-root*, for it is by that name it is known in country places.   I have tested its efficiency in many cases of that common and often fatal disorder, to which young children are subject during the hot summer months in Canada.   For the benefit of anxious mothers I give the following preparation from this valuable root :

Recipe.—Take the long roots, which are covered with a wrinkled brown skin, wash them well and remove the outer bark ; then scrape down the white fibrous part, which is the portion of the root that is to be made use of, throwing aside the inner, hard, central heart, which is not so good.

A large table-spoonful of the scraped root may be boiled in a pint of good milk, till the quantity is reduced to one-half ; a small stick of Cinnamon, and a lump of white sugar, boiled down with the milk improve the flavour, add to its astringent virtue, and make the medicine quite palatable.   The dose for an infant is a tea-spoonful, twice a day ; for an adult, a dessert-spoonful twice or thrice a day, till the disorder is checked.

The months of August and September are the best time to obtain the roots, which have then come to perfection.

The strengthening and purifying nature of this plant makes it quite safe as a medicine even for a young infant. The preparation is by no means unpalatable ; it is sweet and slightly bitter, aromatic and astringent.

I have seen children that had been reduced to the last stage of debility, restored, after taking three or four doses, to a healthy state of body ; it purifies the blood and strengthens the system.

This plant, and *Aralia nudicaulis*, (L.) or Wild Sarsaparilla, are held in great repute as wholesome tonics by the old settlers.

The Ginseng, *A. quinquefolia*, (Gray) or Five-leaved Sarsaparilla, is known by its scarlet berries.

### DWARF GINSENG—*Aralia trifolia*, (Gray.)

Is a pretty, delicate little plant, with three, palmately three to five foliate, light-green leaves, which form a leafy involucre to the small delicate umbel of whitish-green flowers which surmounts them. The root is a round tuber, deep below the soil ; it is pungent to the taste.

### MONKEY FLOWER.—*Mimulus ringens*, (L.)

Our Mimulus is a sober-suited nun, not gorgeously arrayed in crimson and golden sheen, scarlet or orange, but in a modest, unobtrusive dark violet colour, that she may not prove too conspicuous among the herbage and grasses. Her favourite haunt is in damp soil, by low-lying streams and open, swampy meadows, among moisture-loving herbs, coarse grasses and sedges, and dwarf sheltering bushes. Yet our Mimulus is by no means devoid of beauty ; the dark violet-purple of the corollas being rather unusual among wild-flowers. The blossoms grow from between the axils of the leaves, singly, on rather long foot-stalks ; the upper lip of the tubular corolla is arched, the lower spreading and thrice lobed ; the leaves are long, of a dullish green, often, with the angled upright scape, taking a bronzed purple tint.

### MAD-DOG SKULLCAP—*Scutellaria lateriflora*, (L.)

This pretty, light-blue flower grows on the low-lying shores of the Katchawanook Lake, and other localities on the banks of the Otonabee and its tributaries.

The stem is slender, branching, the leaves rather coarse ; colour of the blossoms azure blue ; with the small upper lip somewhat curved.

The old settlers imputed great virtues to this very humble herb, which it is more than doubtful if it possessed. Good faith, however, will often work marvellous cures. The idea was that the plant would avert the terrible effects of the bite of a mad-dog.

E

There is also a much handsomer species with larger flowers and simpler stem—the Common Skull-cap *(S. galericulata.)*

### Marsh Vetchling—Marsh Pea—*Lathyrus palustris*, (L.)

The Marsh-Vetchling or Marsh-Pea is a graceful climbing plant with purple flowers, and long slender leaflets arranged in pairs from two to four, or six, along the leaf-stalk which terminates in a cluster of clasping thread-like tendrils.    The flowers are placed on long, slender, arching peduncles springing from the base of the leaf-stalk, which is furnished at the joint with a pair of sharply pointed stipules.

The Marsh-Pea is found chiefly in damp ground among herbs and dwarf bushes along the margins of low-lying lakes and creeks, and sandy grassy flats.    Its pretty purple pea-shaped blossoms and pale-green leaves attract the eye, as it twines among the herbage and forms graceful garlands amidst the ranker and coarser plants to which it clings.    A taller species with slender stalks, two to four feet high with ovate-elliptical leaves, much larger stipules, and an abundance of small, pale blue-purple flowers is also found on marshy shores.    This is the variety *myrtifolius* of Gray.

There are many other graceful twining plants of this order.    The most remarkable is the

### Indian-bean—*Apios tuberosa*, (Mœnch.)

known also as Indian-potato and Sweet-bean.    A tall climber with compound leaves of five to seven ovate leaflets, and sweet-scented clustered flowers of a brownish-purple colour, and pear-shaped tubers, of the size of a hen's egg, which are used as an article of food by the Indians, who roast them in the embers, and eat them as we do baked potatoes.    A fine white starchy substance can be obtained by grating the tubers—tasteless and not unwholesome.

### Butterfly Weed—*Asclepias tuberosa*, (L.)

Of this remarkable family Canada possesses many handsome species. The most showy is a large bushy plant, with gorgeous orange, almost scarlet flowers.    Every branch is terminated by a wide-spreading head, composed of small umbels of brilliant flowers.    This plant is known by the name of Butterfly Flower from its singularly gay appearance, which is very attractive when seen on dry hills on sunny days.    The root is used in medicine as a powerful vermifuge by the old settlers, who say they learned its medicinal virtue from the Indian herb doctors.

The floral construction of the flowers of all this family is peculiar. The petals are somewhat pointed, five in number ; divisions of the calyx also five ; the petals are reflexed, showing a central crown, which is composed of five hooded nectaries, each of which encloses a curved horn-like appendage. The crown is often of a different shade of colour from the petals ; and from its peculiar form, the flower has the appearance of being double. The leaves of the Butterfly Flower are rough on the surface and hoary ; the seed-pods are also hoary. It is a striking and showy flower, deficient in the viscid milky juice that is so abundant in others of the genus.

The pink-flowered Milkweed, *A. Cornuti*, is fragrant and also handsome ; it is a tall, showy plant, abounding in milky juice ; the leaves are large, soft, and velvety ; the flowers pale pink, falling in graceful tassels from between the leaves ; the form of the flowers is the same as in the above ; the seed-pods are large and the seeds flat, lying one over the other, closely pressed, in beautiful succession, like the shining silvery scales of a fish ; each seed is furnished with a tuft of silken hair.

The pod opens by a long slit ; and it is wonderful to see the beautiful winged seeds, the instant the prison door is opened, rise as if moved by some sudden impulse, spreading their shining silken wings and taking flight, wafted away by the slightest breeze to parts unknown. One marvels how this winged multitude ever found space to lie within the narrow case from which they escaped ; and it reminds you of that wonderful Genius of the old Arabian tale, that the poor scared fisherman induced to re-enter the metal pot. Methinks it would be even harder to gather together our fugitive silky seeds than to coax a refractory Genius into a quart pot again.

The whole of the *Asclepias* family are remarkable for the strong, tough, silken fibre that lines the bark of the stout stem. This, in the common Silk-weed, *A. Cornuti*, has attracted much attention, but has not as yet been utilized for textile fabrics. The fibre is strong, and can be divided into the finest threads of silken softness, and of good length, as the plant reaches from two to three feet, or more, in height, and grows so freely that I have seen extensive plantations of it on wild spots, where it has been self-sown ; and where few other plants would grow.

The silken beard of the seed, though so bright and beautiful, is too short and brittle for spinning ; still, as a felting material, or for paper manufacture, it might prove of value, when even the pod might be employed. A good fibre is found in all the tall Milkweeds and also in the *Apocynums* or Dogbanes, where the thread is still finer. All these plants are remarkable for the bitter, viscid, milky juices with which they abound.

We know nothing in medicine, experimentally, of this tribe of native plants; but I believe they are supposed to contain poisonous properties of a narcotic nature, as is the case with most vegetables containing acrid milky juices.

It would add greatly to the value of botanical books, if a few words as to the poisonous character of native plants were inserted.

### WILLOW-HERB—*Epilobium angustifolium*, (L.)

This handsome, showy plant, with its tall wand-like stem, and abundant blossoms of reddish lilac, adorns old neglected fallow-lands that have been run over by bush fires, and open swampy spots, where it covers the unsightly ground with its bright colours and drooping stems, which are often borne down by the weight of their blossoms and fair buds. It often shares these waste places with the White Everlasting, *Antennaria margaritacea*, Wild Red-Raspberry, Blackberry, and the Fireweed : with a variety of smaller plants that take possession of the virgin soil, there to perfect their flowers and fruit, while at the same time their abundant foliage serves to cover the confusion caused by charred and blackened trunks and branches of prostrate trees. Over all these the graceful Willow-herb waves its flowery spikes and long willowy leaves. All through the months of July, August and September it blooms on, while later in the season its silky-plumed seeds fill the air, as they wing their way to other wild spots equally favourable for their growth and development.

The mid-ribs of the leaves are white, or rosy red, as also are the wand-like stems and branches. The terminal naked buds are of a deep crimson ; the seed-pod long, and opening lengthwise to allow the seeds to float off on the breeze by means of their silky sails.

The Willow-herb is cultivated in gardens in England, where it is known by the name of French Willow. I remember seeing it in almost a wild state, in a picturesque old garden in Suffolk, where it grew to the height of seven or eight feet, the long flowery wand-like stems drooping over the margin of a fish-pond, where, beneath the shadow of a big old Willow, I used to sit and feed the silver-scaled Carp, which were so fearless that they came and fed upon the crumbs that I threw into the water. It was a pleasant spot, with the flowers, and the fish, and the old Willow tree.

EVENING PRIMROSE—*Œnothera biennis* (L.) var. *grandiflora*, (Lindl.)

" A tuft of Evening Primroses
O'er which the mind might hover till it doses,
But that it's ever startled by the leap
Of buds into ripe flowers."—*Keats.*

In common with the Northern and Eastern States, Canada owns many native flowers of this fine family. Our largest variety of *Œ. biennis* is deliciously fragrant, with large showy flowers of a deep sulphur colour —of all the shades of yellow the most beautiful and satisfying to the eye—so full, so soft, and delicate is the hue. Some species of the Evening Primrose, true to their descriptive name, only open their blossoms at sun-set ; others bloom during the day-time and endure the light and heat of a July or August sun. Some, as our variety *grandiflora*, are from three to four feet high, with stout branching stems and many-flowered spikes ; others are low in stature, with rough hoary leaves and smaller flowers. *Œ. pumila*, a dwarf species, about six inches in height, has small flowers of pale colour and of little floral beauty. *Œ. biennis* (L.) var. *muricata*, (Gray), which is common in open fields and plains, is a large branching species with smooth, red-veined leaves, a red bristly stem, and smaller flowers than *grandiflora*. It is less fragrant but is a handsome species, and continues flowering all through the summer, till cut off by early frosts. But by far the finest and most interesting of our Evening Primroses is the large flowered, fragrant, *grandiflora* under consideration : no sooner has the sun set, than one after another may be heard, in quick succession—the bursting of the closely shut sepals of the calyx. One by one the petals begin to unfold—slowly, slowly, you notice a slight movement in the corolla : first one petal is loosened from its plaited folds, then another, till in a few seconds the whole flower expands and opens its beautiful deep sulphur-coloured cup with its eight stamens and yellow anthers, giving out its delightful scent upon the dewy air. What an object of interest is this flower to children as they gaze with watching, wondering, eyes, upon its fair unfolding flowers. One little fellow, almost a baby, cried out, " Oh look ! it's waking now," when he saw the first pure petal softly rolled back as the blossom com menced opening. The diagonal lines which cross the surface of the flower are caused by its twisted æstivation, or folding in the bud, and this gives it a crimped appearance, which is singularly pretty as well as curious. It has been stated that a flash of phosphorescent light has been noticed at the instant the flower opens, but I think a tiny flash of such pale light would hardly be perceptible during the daylight; besides, the petals unclose gradually ; the only sudden motion is the unclasping of the enfolding calyx leaves which emprison the corolla. Nevertheless it is a pretty idea, and it may be a fact, though not as yet a fully established one. I think it is Professor Lindley who has recorded the circumstance in his " Natural System of Botany," from the observation of some French naturalist.

### Enchanter's Night-shade—*Circæa alpina*, (L.)

With so ominous a name we might naturally expect to find some sad, lurid-looking, poisonous weed or sombre-leaved climber, instead of a very delicate, innocent-looking, leafy plant, with thin, light-green foliage, and tiny white or pale pink blossoms, dotted with minute spots of pale yellow, something like the old garden plant London Pride. One can hardly imagine so inoffensive a little flower being introduced by the ancient Sybils into connexion with their unholy rites, nor understand why its classical name, *Circæa*, after a horrible old enchantress, should have been retained by our modern botanists.

We often wonder at the Greek names given to plants which are indigenous to other climes than Greece, and are retained even where the significance is so obscure as to be questioned by our botanical writers. It is these hard classical names that frighten youthful students, especially young ladies, who are only too glad when they can meet with names of flowers that give them an insight into the appearance and qualities of the plants, by which they can be easily recognized.

Imagination loves to get a glimpse at the poetical in the names of flowers, giving a charm to what is dry and uninteresting in our botanical books : something that gives us an insight into the history of the flower we study, beyond the mere structure and definition of its parts. I remember an old gardener (he was by no means an ignorant man) once said, " Oh ! madam, in these days they turn poor Poetry out of doors, but in the olden time it was not so, for it was the language in which God spake to man through the tongues of angels and prophets. Aye, and it was the language in which even sinful man spake in prayer to his Maker : but now they only use hard words for simple things, such as the flowers of the field and the garden; or the talk is about gold, and the things that gold purchases ! "

### Spreading Dogbane [Indian Hemp.] *Apocynum androsæmifolium*, (L.)

This pretty pink-flowered plant is also known by the name of Shrubby Milkweed, from the abundance of acrid milky juice that pervades the stem, branches and leaves.

The flowers of this plant are very unlike those of the *Asclepiadaceæ* ; but it belongs to a closely allied order, and possesses some of the characteristics of that remarkable order of plants in which the deadly

Strychnia is included, with others of evil reputation. There are many virtues as well as vices in our Milkweeds. The Apocynums have some worthy members in the family—sweets as well as bitters.

In the " Hya-hya " of Demarara, we find the luscious Milktree, which with the Cream-fruit of Sierra Leone and some others redeem the character of this remarkable tribe of vegetables. Our own native Shrubby Milkweed has some marked peculiarities which deserve notice ; in common with all the Milkweeds it has a strong, fine, silky fibre in the bark, which can be drawn to a great degree of fineness ; and in one of the species, *Apocynum cannabinum*, Indian Hemp, is exceedingly tough and strong, and is said to have been used by the natives in lieu of thread. No doubt it can be put to such purpose. While many writers have dwelt upon the silk contained in the pods of the Milkweeds, suggesting the possible uses to which it might be applied, the more valuable strong flaxen-fibre, which is superior in quality to hemp, seems in a large measure to have escaped public attention. The free growth of the common white-flowered Milkweed, which could be easily cultivated, growing readily, and attaining the height of three or four feet, would give a long thread easily divided into the finest strands, and might form a valuable addition in the manufacture of native Canadian fabrics. But I have already referred to this subject in another portion of my little work, so I will return again to my text.

The ancient name *Apocynum*, is derived from two Greek words, signifying,—from a dog ; to which this shrub was supposed to be injurious or baneful, whence its common name Dog-bane. Whether the plant deserves this reproach as regards dogs, I cannot say ; but truth obliges me to confess that in its pretty treacherous bells many a poor incautious fly meets with a certain though possibly lingering death. Lured by the fragrance of its blossoms, which it gives out at dew-fall, hundreds of small black flies seek rest and shelter in the flowers, and are seized instantly by the irritable stamens and held in durance by their legs. And as there is no philanthropist to take his nightly rounds and release them, they perish in their flowery prison.

Though the Dog-bane is perennial, the stems die down annually and are renewed again each Spring. The bark is of a deep red ; the foliage on distinct foot-stalks, ovate and pointed. The flowers in loose spreading cymes; the pale rose somewhat striped corolla, open bell-shaped, with recurved lobes. The flowers are followed by long slender red pods, meeting in pairs at the points, in twos or fours, the pods converging together; these pods open longitudinally and let out the small winged seeds, each of which is furnished with a tuft of delicate silk. The whole plant is milky—more so than the next less showy-flowered species

### INDIAN HEMP—*Apocynum cannabinum*, (L.)

The flowers of this species are white, small, and in terminal cymes, the leaves are narrow, of a dark green, smooth; the fibre in the bark of this plant is very strong, as well as fine: the Indians use this thread in the manufacture of fishing nets and lines and probably in sewing. The banks of streams and lakes seem to be the habitat of the Indian Hemp. I am not aware that it has any scent. The scent of the pink Dogbane is only given out after sun-set.

### WHITE DWARF CONVOLVULUS—DAY-FLOWER—*Convolvulus spithamæus*, (Pursh.)

Although so delicate and fragile in texture, there is no flower that loves the sunlight in its noon-tide power more than this lovely wild Convolvulus. In this, it differs from the splendid Morning Glory, which opens early, in the freshness and coolness of the morning, but fades before the noon-day heat and light: only on cool cloudy days will it display its glorious tints of royal purple, rose, crimson, and exquisite shades of pink, pearly-blue, and white. But our modest white flower may be seen blooming in open fallows, and wild grassy plain-lands, where it has little shade unless from the surrounding herbage. The plant is seldom more than twelve or eighteen inches in height, tapering from a broad base to a slender leafy point. The foliage is whitish or hoary grey, from a minute downy covering. These grey leaves are hastate, not arrow-shaped, pointed and lobed at the base; the lower leaves on long foot-stalks, the upper ones diminished to mere bracts. The flowers are large, purely white, open bells, on long stalks—only two opening each day. The stem of the plant is somewhat woody, slightly branching or simple, and forming a pyramid of slender apex, twining slightly and clasping the stalks of grasses and neighbouring herbs.

On the flowery Rice Lake plains, I have seen this lovely flower mingling its hoary foliage and white fragile bells with the gay bracts of the Scarlet Cup and azure-blue spikes of the Wild Lupine, the Sweet Pyrola and Wild Rose: and surely no garden ever shewed more glorious colours or more harmonious contrasts than this wilderness displayed.

This pretty wild Convolvulus might be introduced into garden culture, where the soil is light, without any fear of its becoming a troublesome weed like the common Bind-weed, or the double-blossomed variety, which should only be kept as plants for a Trellis or as Bower-climbers.

Grass-Pink—*Calopogon pulchellus*, (R. Br.)
(PLATE VII.)

Our open, springy Poplar flats, partially shaded by Aspen shrubs and wild grasses, afford shelter to many a rare Orchid. The warm rays of the sun acting on the moist, boggy soil, quicken into life and loveliness one of the most ornamental of our Orchidaceous plants. In the month of July we find that very beautiful flower, the Grass-Pink, or *Calopogon*. Its flowers are little known, and may indeed truly be said to waste their sweetness on the desert air.

From a round, solid corm, about a quarter of an inch in diameter, rises a bright green sword-shaped leaf, which clasps at its base a tall scape, bearing a loose four to eight-flowered raceme of elegant rose or lilac-coloured flowers. The lower blossoms open first. The form of the flower is peculiar ; the concave upper petal or lip is bearded with yellow and purple hairs arching over the column, which is winged and free. The bright reddish-purple sepals and petals are pointed and fragrant ; the scape rises to the height of from eighteen inches to two feet. A bed of these elegant flowers, when in bloom, is a charming sight.

Another of our Orchids is the lovely and rare *Arethusa bulbosa*, (L) the flower of which is no less remarkable for the beauty of its form and rich colouring than the *Calopogon*. The colour of the ringent corolla is of a deep, rich rose-purple, and it is very sweetly scented ; the scape has occasionally one grassy leaf. Not less singular is the charming *Calypso borealis*(Salisb.) or Bird's-foot Orchis, with its graceful, deliciously-scented, pendulous flowers, and crested lip, bearded with yellow and pink, and its narrow, twisted and waved, pale pink sepals and petals ; the scape is garnished with one oval shield-shaped shining leaf of dark glossy green. It flowers in the month of May. Another elegant bog-plant is the

Small Round-leaved Orchis—*Platanthera rotundifolia*, (Rich.)
(PLATE VII.)

" Your voiceless lips, O flowers, are living preacheis ;
Each cup a pulpit, and each leaf a book."
" Floral apostles that in dewy splendour
Weep without woe and blush without a crime."—*Horace Smith.*

This is one of the lovely native plants of the Orchis family, of which we boast many remarkable for beauty as well as for the eccentric forms which arise from the peculiar arrangement of their floral organs.

The one above named is worthy of attention. Our quaint old herbalists would have called it the Holy Dove, or some such name, from the curious resemblance that the petals and sepals take to the body and extended white wings of a hovering Dove. The lower lobed petal taking the semblance of the tail and wings, the upper ones meeting over the anther-cells, which might be likened to the two eyes of the bird, and the arched hooded appendage above, to the head.

The scape of this pretty Orchis is furnished with one handsome round or shield-shaped leaf, of shining bright green, and a bracted spike of white flowers, spotted with delicate pink, as also is the throat of the arched petal that partly covers the anthers and stigmatic disc.

Our beautiful Orchids, with many other rare bog-plants, repay the difficulties of obtaining them in their native haunts, such as Cedar swamps, Cranberry marshes, Poplar swales, and Peat bogs ; where, however zealous, our lady botanists may not venture without risk.

These rare plants, growing in lonely isolated places, are little known and but seldom met with, unless, as I have said, by the enthusiastic botanist who is not afraid to seek for such floral treasures, however difficult they may be to obtain.   A curious and handsome species is the Striped Orchis or Coral-root, *Corallorhiza multiflora*, (Nutt).  This plant is leafless, silvery-sheathing scales taking the place of leaves ; the roots are branched and knobby, like some kinds of coral ; the scapes, many flowered, growing up in clusters from twelve to eighteen inches high ; flowers pale fawn, striped and dotted with crimson or purple ; such was a plant that I found at the root of a big Hemlock tree, near the forest road where I often walked many years ago.

There are several different species of this curious order, varying in size and the colour of their blossoms.

Of fringed and tufted fragrant kinds, we have the Pearly White and the Fringed Pink Orchids.   These are very pretty and not uncommon flowers.   I first saw them on my voyage up the St. Lawrence, when the ship was anchored off Bic Island and the Captain brought me a noble posy of sweet flowers, the first Canadian flowers I ever saw. Among Wild Roses, and elegant Blue Lungwort, *Mertensia maritima*, which I had also seen and gathered near Kirkwall, in Orkney ; there were yellow Loose-strife, Hare-bells, and the sweet scented White Fringed Orchis, the Pink Fringed Orchis and some elegant cream-coloured Vetches, with several other flowers then unknown to me.

There are many other plants of the Orchis family scattered through our woods and swamps, and on the rocky or low islands of our Northern Lakes.  Among those not already mentioned, the Larger Fringed Orchis *Habenaria fimbriata*, may be named.   This is a tall handsome bog-plant, flowering in the beginning of July, with large rose-purple, deeply-cut petals.   Another less conspicuous species, found in dry woods, is the Northern Green-man Orchis, *Habenaria viridis* (L.) var. *bracteata,*

(Reich.) The scape of this species is furnished with long narrow sharply pointed bracts and greenish flowers.

In some of our Orchidaceous plants when examined, there will be seen at the base of the fleshy scape, two roundish bulbs, or tubers ; farinaceous masses, whence the bundle of white fibres, the roots and rootlets proper, proceed, and which contain the prepared food to support the growth of the year.

From one of these tubers, the scape, bearing the scaly or leafy bracts, root-leaves and flowers, springs, and at the flowering season is much larger than the other.

The flower-bearing bulb deceases from exhaustion of its substance, shrivels, turns brown, and begins to decay, while the other continues slowly but steadily to go on increasing, bearing in its bosom the embryo flower-stem and foliage, which are to appear the following year. Another tiny bulb is also preparing in like manner, attached by a slender fleshy cord to its companion. Thus from year to year the process goes, on each one taking the place of its predecessor after its office has been fulfilled.

This singular mode of reproduction seems to supersede the necessity for the development of seed, as in other flowering plants ; nor is it so common to find seedlings of the Orchids springing up round the parent plant, as in the case of other flowers.

The reason why so few amateur florists succeed in transplanting the native Orchids into their gardens, arises from want of due care in taking them up. The life of the plant for the following season being contained in the new forming tuber, if this be in the least injured the chance of another flower in the future is at an end. The succulent tender roots are easily broken or wounded, and these strike rather deep down in the soil, and must be taken up uninjured, with a good portion of the mould, or there is small chance of life for the plant. Nor will the Orchis thrive in common earth—it requires fibrous, peaty soil, moisture, and some shade, with the warmth that arises from the moist soil, and shelter of the surrounding herbage. They all thrive best in the Conservatory or Green-house.

GOLDEN DODDER—*Cuscuta Gronovii,* (Willd.)

This singular parasitical plant occurs on the rocky shores of our inland lakes. There seem to be two species. One with bright, orange-coloured coils, and greenish white flowers ; the other with green, rusty wiry stems, and smaller blossoms. This last occurs on the rocky shores of Stoney Lake, where in the month of August it may be found twining around the slender stems of the Lesser Golden-rod, a small, narrow-leaved Solidago.

In no instance did I find this curious parasite associated with any other plant; as if by some mysterious instinct the Golden-rod seemed to be selected for its support. Nor could the union with the flower be discovered by the most careful examination. The Dodder seems to be leafless and rootless. The Golden-rod to which it had attached itself did not appear to have suffered from the clinging embrace of its singular companion, though its coils were so tightly wound around it that it was not an easy matter to separate them from the supporting stem.

The Dodder could not even be said to have the claims of a poor relation to excuse its unwelcome intrusion.

The white blossoms of this parasite were closely clustered in intervals on the wiry stem.

The Golden-stemmed species, with somewhat larger, greenish-tinged white flowers, I found in the same locality attached to the culms of stout wild grasses, which alone it seemed to have selected for its support. The bright orange coils, and clusters of flowers, formed a pretty contrast with the dark foliage of the climbing Indian Bean, *Apios tuberosa*, many young plants of which handsome, fragrant climber grew there in profusion, covering the low bushes.

In the States it is known as Gold Thread, from the bright, orange, thready twining stems, which it throws like a golden net over the neighbouring herbage. It seems, indeed, more ornamental than useful ; but as it does not intrude itself into our gardens, we will not quarrel with it. There is room and space in this wide world for it and others to find some little spot in which to grow. Something would miss it, were it to be entirely destroyed from the face of the earth—for as the poet says—

> " Nothing lives, or grows, or moves in vain ;
> Thy praise is heard amid her pathless ways,
> And e'en her senseless things in THEE rejoice."—*J. Roscoe.*

### EVERLASTING FLOWERS.

> " Bring flowers for the brow of the early dead."

It is on the open prairie-like tracts of rolling land, known in Ontario by the names of Oak-openings and Plains, where the soil is sandy or light-loam, that flowers of the Composite order abound. All through the hot months of July and August and late into September, the starry rayed blossoms of the sun-loving Sunflowers, Rudbeckias, Asters and Golden-rods, enliven the open wastes and grassy thickets, with their gay colours, the more welcome because the more delicate of the early Spring and Summer flowers have long since faded and gone, and we know that we shall see them no more.

Our Floral Calendar might be likened to four stages of life. The tender early flowers of Spring to innocent childhood. The gay blossoms of May and June with all their fruitful promises, to advancing youth. The ripening fruit of Summer's prime, to mature manhood in its strength and perfection ; while the white flowers and hoary leaves of our Pearly Everlastings and drooping Grasses are not inapt emblems of old age, bending earthward yet not destroyed, for they have winged seeds that rise and float upwards and heavenwards, and we shall again behold them in renewed youth and beauty.

EARLY-FLOWERING EVERLASTING—*Antennaria dioica*, (Gaertn.)

Our earliest Everlasting is a pretty, low, creeping plant, not exceeding six inches in height, with small round clustered heads of downy whiteness, with dark brown anthers, which resemble the antennæ of some small insect, whence the generic name *Antennaria* is taken. The leaves of the plant are white beneath and slightly cottony on the outer surface, becoming darker green during the summer. The root-stock is spreading, the leaves numerous, roundish-spatulate. The whole plant has a hoary appearance when it first springs up.

This modest, innocent-looking little flower, peeps forth in April, and carpets the dry, gravelly hills with its downy blossoms and soft, silken leaves, sharing the newly uncovered earth with the Blue Violet (*Viola cucullata*), and early pale yellow Crowfoot, Rock Saxifrage and Barren Wild Strawberry (*Waldsteinia fragarioides*, (Tratt), which is then beginning to put forth its new foliage and yellow flowers, that have been kindly sheltered by the persistent leaves of the former year, now red and bronzed by the frosts of early Spring. Our pretty Canadian Everlasting bears some family resemblance to the far-famed "Edelweiss" of the High Alps (*Leontopodium alpinum*). As in that flower, the clustered heads are set round the centre of the disc, like a little infant family surrounding the careful mother.

In the singular Alpine species, the whole plant, from root-leaves to stem and involucre, is thickly clothed with snow-white down, as if to keep it warmly defended from the bitter mountain blasts and whirling showers of snow and hail. Thus does Creative Love shield and clothe the flowers of the field : His tender care is over all His works.

Scarcely has our little Everlasting raised its soft cottony head above the short turf when another species appears, as if to rival its tiny brother.

PLANTAIN-LEAVED EVERLASTING—*Antennaria plantaginifolia*, (R. Br.)

This plant varies in height from six inches to eight or nine. The woolly stem is clothed with narrow, leafy bracts ; the root-leaves are large and broadly ovate, several-nerved, very white underneath, and less downy on the outer surface ; the corymbed head of flowers shining with bright scales and silky pappus—the scales are not pure white, but with a slight tinge of brown.   Later on in the month of July, a tall, slender form of this Everlasting may be seen with larger root-leaves and loose heads of flowers on long foot-stalks ; the flowers are slightly tinged with reddish-purple and silvery-grey, which gives a pearly or prismatic effect, as the eye glances over a number of the plants moved by the summer wind.   The flowery heads are conical, the unopened blossoms sharply pointed : the whole plant tall, slender and simple, and very downy.

The later plants of the Everlasting family differ from the above species.   One commonly called

### Neglected Everlasting—*Gnaphalium polycephalum*, (Mx.)

deserves our especial notice on account of the pleasant fragrance which pervades the gummy leaves, as well as the shining straw-coloured flowers; the scent is aromatic and slightly resinous.   This plant is found in old pastures, and by wayside waste lands, often mingled with the Pearly-Everlasting (*Antennaria margaritacea*) and other common species of the order.

It is so commonly seen, and is so little cared for as to have obtained the name of Neglected Everlasting.   Truly even a flower may be without honour in its own country!

There is another plant of this family, found in old dry pastures, with straw-coloured, shining flowers ; but it lacks the aromatic fragrance and dark-green, narrow, revolute, gummy leaves of the preceding ; it is branching with a wide-spread corymbed head and has the leaves decurrent on the stem, whence its name, *G. decurrens.*   This is an earlier species than the Neglected Everlasting.

### Pearly Everlasting—*Antennaria margaritacea*, (R. Br.)

The abundance of the common Pearly Everlasting induced many of the backwoods settlers' wives to employ the light dry flowers as a substitute for feathers in stuffing beds and cushions, and very sweet and comfortable these primitive pillows and cushions are, they are, too, pleasantly fragrant, for the Pearly Everlasting is also sweet-scented, though not so much so as *G. polycephalum* ; the heads are soft, elastic, and easily obtained.   The French peasants still hang up Wreaths or Crosses of the white-flowered Everlastings in churches, and upon the graves of the dead, to mark where one fair bud or blossom has dropped from the parent tree to mingle with its kindred dust. It is a fond old custom, which time and the world's later fashions have not yet changed among the simple *habitants.*

Surely we may say with the sweet poet :—

"They are loves last gift,
Bring flowers—pale flowers."

### YELLOW COLTSFOOT— *Tussilago Farfara*, (L.)

A large proportion of our flowers of Mid-summer and Autumn are of the Composite order, but in the Spring they are rare, with a few exceptions, such as the Early-flowering Everlasting, the Fleabanes and the Coltsfoot.

The first flower that blossoms is the Coltsfoot, *Tussilago farfara*, (L.) which breaks the ground in April with its scaly, leafless stem and single-headed, orange-yellow rayed flower. It is a coarse, uninteresting plant, not common, excepting in wet, clayey soil ; seldom found in the forest. It is the earliest plant of the Canadian Spring, and prized on that account, and for its medicinal virtue, real or imaginary. Both flower and leaf are larger than the British species, but its habits are similar.

In July, August and September our rayed flowers predominate, especially in the two latter months ; it is then, when the more delicate herbaceous flowers are perfecting their seeds, that our hardy Sun flowers lift up their showy heads and seem to court the glare of the summer sunshine ; it is then that we see our open fields gay with Rudbeckias, Chrysanthemums, Ragworts, Golden-rods, Thistles and Hawk-weeds. In the forest we find our White Eupatoriums, Prenanthes, and Fire-weeds. On all wastes and neglected spots the wild Chamomile abounds, as if to supply a tonic for all agues and intermittents. The beautiful Aster family may now be seen in fields, by waysides, on lonely lake shores, in thickets, on the margins of pools and mill-dams or waving its graceful flowery branches, on the grassy plains and within the precincts of the forest. There are species for each locality—white, blue, purple, lilac, pearly-blue—with many varieties of shade, height and foliage ; some species graceful, bending, and spreading, others stiff, upright and coarse ; but the species are number-less and their habits as various. The most elegant are the *Aster cordi-folius*, (L.) and *A. puniceus*, (Ait). The most delicate, the little white, shrubby Aster, *A. multiflorus*, (L.) with reddish disc and golden-tipped anthers, which give a lovely look to the crowded, small, white-rayed flowers, as if they were spangled with gold-dust. On dry, gravelly banks,

near lakes and streams, is the favourite haunt of this pretty Aster.    The
plant is much branched, the branches growing at right-angles to the stem ;
crossed with narrow leaves, and bearing an abundance of small, daisy-like
blossoms.    On the springy shores of ponds and the banks of low creeks
ın upright, single-headed Aster, *A. æstivus*, may be seen, with bright azure
rays and yellow disc, together with a tall   woody-stemmed   flat-topped,
coarsely-rayed, white species, *Diplopappus umbellatus*, (T. & G.).    The
large-flowered, branching, many-blossomed, purple-rayed Asters are chiefly
found in dry fields, by wayside fences, and among loose rocks and stones,
giving beauty where all else is rough and unsightly, making the desert to
blossom as a garden ; so bountiful is Nature ; so beneficent is the Creator
in all His works ; so lavishly does He scatter man's path with flowers, that
his eye may be gladdened and his heart may rejoice in the beautiful things
of the earth on which he is a sojourner.    Should we not, therefore, praise
Him  even  for the lowly herbs and the  lovely  blossoms that adorn our
paths.

CONE-FLOWER—Rudbeckia hirta, (L.)

The Cone-flower is one of the handsomest of our rayed flowers.
The gorgeous flaming orange dress, with the deep purple disc of almost
metallic lustre, is one of the ornaments of all our wild  open prarie-like
plains during the hot months of July, August and September.   We find
the Cone-flower on  sunny spots among the  wild herbage of grassy
thickets, associated with  wild Sunflowers, Asters and  other  plants of
the widely diffused Composite Order.

Many of these compound flowers possess medicinal qualities.  Some,
as the Sow-thistle, Dandelion, Wild Lettuce,  and  others, are  narcotic,
being supplied with an abundance of  bitter milky juice.   The Sunflower,
Coreopsis, Cone-flower, Ragweed, and Tansy, contain resinous properties.

The beautiful Aster family, if not remarkable for any peculiarly
useful qualities, contains many highly ornamental plants.   Numerous
species of these charming flowers belong to our Canadian flora ; lingering with us

"When fairer flowers are all decayed,"

brightening the waste places and banks of lakes and lonely streams
with starry flowers of every hue and shade—white, pearly-blue and
deep purple.

The Cone-flower is from one to three feet in height, the stem simple,
or branching, each branchlet terminating in a single head.   The rays are
of a deep orange colour, varying to yellow ; the leaves broadly lanceolate,
sometimes once or twice lobed, partly clasping the rough, hairy stem,

hoary and of a dull green, few and scattered. The scales of the chaffy disc are of a dark, shining purple, forming a somewhat depressed cone. This species, with a slenderer-stemmed variety, with rays of a golden yellow, are to be met with largely diffused over the Province.

Many splendid species of the Cone-flower are to be found on the wide-spread prairies of the West, where their brilliant starry flowers are mingled with many a gay blossom known only to the wild Indian hunter, and the herb-seeking Medicine-men of the native tribes, who know their medicinal and healing qualities, if they are insensible to their outward beauty. One tall, purple-rayed species (*Echinacea purpurea*) is very handsome.

I sometimes think, that though apparently indifferent to the beauties of nature, our labourers are not really so unobservant or apathetic as we suppose them to be ; but that being unable to express themselves in suitable language, they are silent on subjects concerning which more enlarged minds can speak eloquently, having words at their command. The uneducated know little of the art of word painting, in describing the beautiful or the sublime.

SPICE WINTER-GREEN—*Gaultheria procumbens* (L).

This pretty little plant has many names besides the one above : it is also known as Tea-berry, Checker-berry and Aromatic Winter-green ; but it shares these English names with many other forest plants.

The aromatic flavour of its leaves and berries has made the Spice Winter-green a favourite, not with the Indians only, but also with the Confectioners, who introduce the essential oil that is extracted from the leaves and fruit into their sugar confections. It is also an ingredient in many of the tonic and alterative bitters prepared and sold by the Druggists in Canada. The Squaws chew the dry, spicy, mealy berries when ripe with great relish ; and in the lodge, the Indian hunter smokes the leaves as a substitute for tobacco : when burnt they give out a pleasant aromatic smell. The leaves are warm and stimulant, agreeable to the taste, and perfectly wholesome.

The creeping root-stock throws up simple upright stems at intervals, crowned with a few smooth, thick, shining leaves, of a bright green colour. The flowers are three or four in number, resembling in form the Arbutus, Heath, Huckle-berry and others of the family ; being a roundish bell contracted at the neck, pale-white or flesh-coloured. The fruit which is persistent through the winter, is of a brilliant scarlet. The fleshy calyx is of the same texture and colour, and forms a part of the

F

edible berry. The habit of the plant is evergreen, and it may be found on sandy knolls, in thickets, and under the shade of bushes in Oak-openings; a finer, larger form is also to be met with in the forest, in cedar swamps; the leaves, fruit and flowers being nearly twice the size of the above. The leaves are strongly revolute at the edges, very smooth and shining.

There is nothing that we cling to with fonder affection than the flowers of our country, especially such as in childhood we delighted to gather. Thus the Daisy, Primrose and Violet of England and Ireland, and the Bonnie Heather and Harebell of old Scotia, are dear to the heart of the emigrant, and the sight of one of these beloved flowers cherished in a garden or green-house, will awaken the tenderest emotions. An old Scotch woman when asked how she liked Canada, replied. "Aye, nae dout its a gude land for food, and for the bairns, but there is nae a bit of heather, or ae bonny Bluebell in a' the lan'. Its nae like my ain country."

When shown a bunch of Harebells, which I had gathered fresh from a gravelly bank; she grat (wept) at the sight of them. "To see," she said, " the bonnie wee things once mair before I died."

I was once touched by the rapture, even to tears, of a Swiss nurse, who on seeing some flowers of the Alpine Ranunculus growing in the garden of Tavistock Square, flung herself on the grass beside them and kissing each blossom, cried out, " Ah! flore de ma pays." " Ah! flower of my country."

The brilliant scarlet berries of several of the shrubby little Winter-greens, forming so gay a contrast to the dark, glossy foliage, render them very attractive.

On dry rocky hills we find the Box-leaved Winter-green or Bearberry, *Arctostaphylos Uva-ursi*, (Spreng.) which clothes the dry, rocky and gravelly hills all through the continent of North America, is found far to the North, even in barren Labrador, and on the rocky slopes of the far-off Hudson's Bay. It abounds far north in Norway, and clothes the ground with its spreading branches. As winter approaches the dark green leaves assume a purplish-bronze hue which is enlivened by the bright red berries. These pretty evergreens might be adopted as a substitute for the Holly, by such as care to keep up the old custom of dressing the house with green boughs at Christmas-tide in honour of the birthday of the Saviour. Might not the primitive Christians have intended by these emblems to keep Faith, Hope and Charity ever green within the Church and Homestead.

A deeper meaning often lies in the old usages of our forefathers than we are willing to acknowledge in this our day of cotton-spinning and gold-digging, railroads and electric telegraphs.

### RATTLESNAKE ROOT—*Nabalus albus*, (Hook.)

This tall stately-growing plant belongs to the same Natural Order as the Lettuce, and like it abounds in a bitter milky juice, which pervades the leaves and stem and thick spindle-shaped root, even to the pedicels of the graceful nodding pendent flowers.

The plant applied both externally and internally has long had the reputation of being an antidote for the bite of the Rattlesnake.

The slender ligulate .corollas which surround the cinnamon-coloured pappus, are beautifully striped with purple and creamy white ; the pointed tips are turned backwards in the full-blown flowers displaying the stamens and pistils, and soft woolly pappus. The clustered flowers on slender foot-stalks, droop very gracefully at intervals on the stem, which with the branchlets have a purplish tinge.

In the variety *Serpentaria*, this colour pervades the whole plant to a greater degree, and the leaves are more deeply divided than in the type.

In damp rich woods we often find a slender delicate species, which is commonly called

### LION'S-FOOT—*Nabalus altissimus*, (Hook.)

The plant is from two to three feet high ; leaves light green, thin, coarsely toothed and widely lobed. The strap-shaped flowers, narrow, pointed and revolute ; the scales are of a pale green, the pappus of a beautiful fawn colour. The elegant yellow drooping flowers in clusters, making this forest plant a very attractive object from its graceful habit

The above plant was pointed out to me as the *true* Lion's-foot, by an old Yankee settler, and I have retained the name, though it does not quite correspond with Gray's plant, so called. Gray's Lion's-foot is also known as Gall of the Earth, from the intense bitterness of its root ; possibly all these bitter milky juiced plants are narcotics, but as yet not recognized unless by the unlearned Indian, or old herbalist of some remote backwoods settlement, where doctors and druggists were unknown, and the herbs of the field the only medicaments ; generally administered by an old woman, famed more for her herb decoctions and plasters than for her wisdom in book-learning, who believed that there was a salve for every sore, and a potion for every ailment under the sun if the folk had but faith to believe in her "Yarbs."

### THOROUGH-WORTS.

There is a popular belief among many of our native Herbalists, that for every disease that man is subject to, God in His mercy has provided a certain remedy in the herbs of the field and trees of the forest ; that there is a sovereign virtue in roots and barks, and leaves and flowers, if man will but search them out and test their qualities.

The use of simples, as the vegetable medicament used emphatically to be termed, has always found advocates in the lower classes, especially amongst the humble country-folk, who dread mineral medicines, with the nature of which they are totally unacquainted—preferring the herbs of the field, which they see growing about them, to the more costly Doctor's stuff, as they call the prescriptive medicines of the physician. To the Herb Doctor they apply with every confidence, entertaining no fear of the vegetable poisons, in which he often deals ; in his skill they have unlimited faith.

Much of this kind of knowledge is possessed by the old Canadian and Yankee settlers, those hardy pioneers who emigrated from the United States at the close of the Revolutionary War, induced by the promised reward of certain grants of land in return for their professed, or actually proved attachment to the British Government. These families under the appellation of U. E. or United Empire Loyalists, spread themselves along the then unbroken forests on the shores of the St. Lawrence, and bore hardships and privations, to which there are few parallel cases.

Dwellers in the lonely, leafy wilderness, with no road but the rushing river, or broad-spread sea-like lake. They lived apart from their fellow-men : self-dependent, they relied upon their own ingenuity and personal exertions for the actual necessaries of life. The men supplied the household with game from the forest (it was over-plentiful in those days) and fish from the lakes and streams. While in clearing the land and cultivating it in the rude fashion of those days, the women and children, without respect of age and sex, did their part. On the females depended the manufacture of every article of clothing ; the loom occupied a prominent place in the log-house, and the big spinning-wheel the stoop in Summer.

Occasionally a few families bound together by ties of love, or interest, wisely formed a colony and lived within a reasonable distance from one another ; but more commonly their grants comprising many hundreds of acres, according to the number of persons in one household, the settlers were thrown far apart. A blazed path through the forest their only means of communication by land ; and this often interrupted by rapid unbridged streams, or impenetrable cedar-swamps.

In case of accidents, such as wounds from axes, broken limbs, and such ailments as agues and fevers, necessity compelled active measures to be adopted on the spot ; medical-practitioners, so called, there were none ; the broken limbs were set by those in the settlement possessed of the most nerve, while the elder women bound up the wounds, or gathered the healing herbs which they had learned to distinguish by experience, or from oral tradition, as being curative in certain disorders. Something of this healing art was derived from their ancestors, who had the knowledge from the Indian medicine-men ; and some remedies were no doubt discovered by chance ; a happy thought seized upon, and put into practice in some desperate case, where the chances of life hung upon something being done to relieve the sufferer, effected a cure, and established the fame of the remedy.

To these simple people, no doubt, we owe many of the significant local names by which our native plants are still distinguished, and which will always be adopted when speaking of them in familiar parlance. Occasionally we pause and ponder on the source whence such a name as Boneset for *Eupatorum perfoliatum* (L.) has been derived. We can only surmise that the powerful virtues of the plant are serviceable in cases of dislocations and fractures, by reducing fever and causing a more healthy action of the blood, which accelerates the return to strength in the injured limb.

The sanative qualities of these plants are no new discovery, nor are the medicinal properties confined to one species alone, some are used in curing the bites of snakes, as *E. ageratoides* (L.) An infusion of the leaves of another species is an excellent diet drink, almost all are sudorifics and tonics.

The genus Eupatorium is dedicated to Eupator Mithridates, who is said to have used a species of the genus in medicine. Several species of these homely plants are used in Fevers and Intermittents by the Herb-doctors and Indians.

The tallest and most showy of the Eupatoriums is

TRUMPET-WEED—THOROUGH-WORT.—*E. purpureum,* (L.)

The flowers, in dense corymbs are of a deep flesh-colour, approaching to red ; leaves, shining, coarsely veined, narrowing to a point, the upper ones much narrower, mostly growing in whorls round the stout stem. The plant has a bitter, somewhat resinous scent when the leaves are bruised. This tall Thorough-wort is abundant on the banks of creeks and in marshy places, where it often reaches the height of five or six feet.

The red-flowered Eupatorium, the old Thorough-wort of the English herbalists, seems to resemble our Canadian plant very closely ;

its habits, colours and qualities seem the same. When viewing the native species it seems to carry my thoughts back to childish haunts on the banks of the clear-flowing Waveney and the flowery Suffolk meadows

> " Where in childhood I strayed,
> And plucked the wild flowers that hung over its way. "

A more graceful member of the Eupatorium family is the

WHITE SNAKE-ROOT—*Eupatorium ageratoides,* (L.)

which is a pretty, elegant, shrubby plant found in rich woods. The white flowers are borne in compound corymbs.

The leaves are from two to three inches long, toothed, narrowly pointed, on long stalks and of a bright green, smooth and thin. Our plant is about three feet high, wide and loosely spreading. The pretty white corymbs of flowers make this an attraction, seen among the forest herbage ; for at the season when it is in bloom most of the flowers have disappeared from the woods. Not unfrequently we find in damp woods, but more especially on open marshy ground, the well-known herb

BONE-SET—*Eupatorium perfoliatum,* (L.)

This species is easily distinguished from any other by its veiny, hoary, greyish-green leaves, united at the base around the stem, or perfoliate, the stem of the plant passing through the centre of each pair. The large, closely-set corymbs of flowers are of a greenish-white and want the pretty tasselled appearance of the White Snake-root, *E. ageratoides.* The scent of this more homely plant is strongly resinous and bitter ; but it is held in great esteem for certain qualities of a tonic and anti-febrile nature, and forms one of the old remedies for ague and fever.

In evidence of the value of the herb Bone-set, Pursh gives a practical illustration from his personal experience of the efficacy of its medicinal virtues. He says :—

" The whole plant is exceedingly bitter, and has been used for ages past by the natives in Intermittent Fevers ; it is known by its common names, Thorough-wort and Bone-set. During my stay in the neighbourhood of Ontario, when both Influenza and Lake-fever were raging, I saw the benefit arising from the use of it, both as regarded myself and others It is used as a decoction, or as I considered more effectual, as an infusion or extract in rum or gin." ( *Vide* Pursh's *Flora Americæ Septentrionalis ).*

## May–Weed—*Maruta Cotula*, (DC.)

The traveller passes by
With reckless glance, and careless tread,
Nor marks the kindly carpet spread.
Beneath his thankless feet.

So poor a meed of sympathy,
Do gracious herbs of low degree,
From haughty mortals meet.—*Agnes Strickland.*

This is one of our commonest weeds, intruding itself into the very streets and by-lanes of our villages, but never welcome there, as it gives out a nauseous bitter scent at dew-fall. The more sunny the place and the drier the soil, the more does this hardy plant flourish; it heeds not the trampling feet of man or steed, but rises uninjured from the tread of the passers-by, cheerful under all persecution, despised and disregarded as it is; yet if we look closely we see beauty in the finely cut and divided foliage, and the ivory-white, daisy-rayed flowers which appear all through the summer; but when seen in dirty streets we overlook its merits and turn from it with distaste. This feeling is not very amiable, but it is natural, to dislike whatever is vulgar, low and intrusive.

## Wild Sun–flower—*Helianthus strigosus*, (L.)

" As the Sun-flower turns to her god as he sets,
The same look which she turned when he rose."—*Moore.*

So sings the Irish bard, but I rather fancy it is a poetical illusion, for I have watched the flowers, and never could convince myself of the fact. However we may hope that as the Sun-flower has become so fashionable an ornament in the present day, some of its devoted lovers will strive to ascertain the truth of the tradition.

As a not very graceful badge of the votaries of æstheticism, we see the garish orange Sun-flower worn in hats and bonnets, as ornaments for breast and sleeves; and reproduced in needle-work and other ornamental designs for the boudoir or drawing-room. Rows of the gigantic flowers may now be seen lolling their jolly heads in gardens, and lording it over the humbler and lowlier blossoms. Tastes differ—-I am afraid my wild Sun-flowers would hardly be appreciated by some of the fashionable ladies and gentlemen, followers of Oscar Wilde.

We have many flowers of this wide-spread tribe of plants extending through the Country, wherever the soil and surroundings are favourable to their growth; especially may different members of these rayed flowers be found on dry plains, in open copse-woods, and on the banks of streams where the soil is sandy, or gravelly.

So numerous are the varieties, that it would be tedious to enumerate them. One of the handsomest is *H. strgiosus*, (L.) The Sun-flowers form one of the distinguishing floral ornaments of the Canadian plains, and of the extensive prairies of the North-West, where miles of Sun-flowers, Rudbeckias, Liatris and other gorgeous flowers—blue, white, red—may be seen all through the hot summer months. The orange and yellow stars of the Helianthus tribe above all most conspicuously apparent.

The garden Sun-flower may often be met with within the forest, the seed having been carried by the Ground-hog or Squirrel, and dropped on the road.

I have seen little piles of the ripe seed of the garden Sun-flower lying on stumps and rails to dry ; the industrious little gleaners depositing them in such places, to be hoarded at their convenience in their granaries. The same thing may be noticed during the harvest-time near the Wheat-fields.

I have watched with no little curiosity, the heaps of Wheat left by these little, innocent gleaners, and seen them come with their companions to fetch away their newly threshed stores, having first carefully destroyed the germs. Who taught the Squirrel that wise precaution, to prevent the germination of the grain ?

Many years ago, while living on a wild lot, on the Rice Lake, my son in digging the ground for the construction of a root-house, discovered a granary of a Squirrel, or it might be of a Ground-hog, the Canadian Marmot. A large supply of Indian Corn, Beech-nuts and Acorns, was stored many feet below the surface of the dry sandy soil ; but the eye or germ had been carefully bitten out of each one.

DANDELION—*Taraxacum Dens-leonis*, (Desf.)

The Composite Order presents us with more numerous families of plants than any other, and supplies us with a host of flowers, and also some troublesome weeds, which are of wide diffusion, the winged seeds being borne to great distances, and establishing themselves wherever they chance to alight. Many an un-named flower exists, no doubt, in unfrequented spots, where as yet the foot of man has never trod,—those unfrequented wilds where even the hardy lumberman's axe has never been heard, those rugged hills known only to the Eagle and the Falcon ; those deep cedar swamps that afford shelter to the Wolf, the Bear and the Wild-cat, conceal many a graceful shrub and rare plant that one day may be gazed on with admiring eyes by the fortunate naturalist whose reward may possibly be to have his name conferred upon the newly discovered floral treasure.

A large number of plants of the Composite Order are remarkable for the bitter milky juice contained in the leaves, stalks and roots, the properties of which are narcotic and sedative. This bitter milky juice pervades all parts of the Dandelion, or *Taraxacum* ; also Wild Endive and other members of the Lettuce tribe.

The Dandelion is so well known that it is unnecessary to enter into any description of its floral parts. The root of the Dandelion has been utilized as a substitute for Coffee ; in preparing it the root should be washed thoroughly, but the thin brown skin not scraped off, as much of the tonic virtue is contained in this brown covering of the root. This must be cut up into small pieces and dried by degrees in the oven until it becomes dry and crisp enough to grind in the coffee-mill ; it is then used in the same way as the Coffee-berry, with the addition of milk and sugar. A small portion of fresh Coffee would, I think, be an improvement to the beverage, but it is not usually added. Many persons have used this preparation of the Dandelion and greatly approved of it. It is a good tonic and very wholesome. The herb itself, if the leaves be blanched, makes a good salad, equal to the garden Endive.

### PURSLANE—*Portulaca oleracea*, (L.)

This is one of the troublesome weeds of our gardens, and one would hardly associate it with the brilliant, showy flower of our borders. We must, however, recognize it as a near relation. The original of the cultivated Portulaca of our gardens is *P. grandiflora*, from South America, whence it was introduced some years ago. Even in its wild state, or on its native prairies, it is a strikingly attractive flower claiming the admiration of the beholder, but our humbler species is regarded as a thing of naught. The simple Purslane however, has its virtues, and we will try to rescue it from being utterly despised, by showing how it may be utilized. When the plant first appears it pushes forth small wedge-shaped succulent leaves, of a dull red colour, and soon spreads over the ground, branching at every thickened joint. If the soil be rich it becomes very luxuriant, and being very tenacious of life it is difficult to get rid of it, as it springs again from the joints, flourishing the more vigourously from the persecution it has undergone. The axil of every joint is furnished with a small sharply-pointed red bud. The flowers are small, pale yellow, opening in sunshine ; pod, many seeded, with a little round lid that covers the top of the capsule.

The soft, oily mildness, of the leaves and stalks of this plant, renders it useful as an application, crushed or steeped in hot water or milk, for inflammatory tumours. I have seen it also recommended as a pot-herb for the table—in fact, it is largely grown in France for that purpose ; I have also heard it said that it may be used as a dye, but that the blue colour produced is very evanescent.

I merely mention this about the use made of this plant as a dye weed, but have no experience of my own to verify its accuracy.

### WILD BERGAMOT—*Monarda fistulosa*, (L.)

Among the Mints we have many different species, all odorous, pungent and aromatic ; some have pretty flowers, but generally speaking they are more valued for their qualities than chosen for any striking beauty of colour in the blossoms.

We have Spear-mint, Pepper-mint, Horse-mint, Catnip and many others of this humble but not useless family.

The plants of the Natural Order Labiatæ are remarkable for being mostly aromatic and pungent, although some are coarse and rank in odour, none are hurtful.

One of the handsomest and most agreeable in scent is the tall *Monarda* or Wild Bergamot, a very handsome, sweet-scented plant, common upon our Oak-openings and wild grassy plains and dry uplands. I have seen a very pretty variety—*Monarda fistulosa* (L.) var. *mollis*, (Benth)—with rose-coloured blossoms and glandular flowers, from the Poplar Hills, Manitoba. The species so commonly seen on the hilly ground above Rice Lake—*Monarda fistulosa*, (L.)—is tall, with soft leaves of a dull green, of a fine aromatic scent and velvety surface ; the globular heads of lilac lipped flowers are terminal ; the colour of the corolla varies from lilac to very pale, pinkish-white.

All the species are sweet-scented, and might be utilized to advantage as an aromatic flavouring. The Bergamot being far more delicate and agreeable than the Winter-green which is so largely used in confections.

### HEAL-ALL—*Prunella vulgaris*, (L.)

This simple herb is commonly found in grassy meadows and on wayside waste-lands, near rivers and low grounds. It is common everywhere, yet is generally thought to be an exotic, having been introduced among foreign grasses, and thus become naturalized to the country.

There seems to be really no special virtue in the plant ; though it boasts of a name which should entitle it to notice, yet we are ignorant of its medicinal or healing uses. It is destitute of any sweetness, but the blossoms are pretty and associated with English meadows and green bowery lanes, so we look kindly upon the purple-lipped flower for the dear Old Country's sake.

## COMMON MULLEIN.—*Verbascum Thapsus*, (L.)

This plant is one of the tallest of our wayside weeds ; the large, soft leaves, densely clothed with silky white hairs, are not without value with the Herb-doctors. They are used in pulmonary disorders, as outward applications for healing purposes, and in such complaints as Dysentery, to allay pain ; the leaves are made hot before the fire, and so laid over the body of the sufferer. Moreover, this wonderful plant is said to drive away rats and mice, if laid in cellars or granaries ; but this virtue may only be a fond delusion. Commend me rather to Miss Pussy, as a more certain exterminator of these troublesome household pests. A grand and stately spike of golden flowers, called Giant-taper, grew in my father's garden, and was the resort of Honey-bees innumerable. Homely as our Canadian plant is considered to be, yet it has uses of its own, besides those attributed to it by the old settlers. The abundance of the seeds which remain in the hard capsules during the winter, afford a bountiful supply of food for the small birds that come to us early in Spring. In March, and early in April, the Snow-Sparrows, and their associates, the little Chesnut-crowned Sparrows,

" That come before the Swallow dares,"

and the brown Song Sparrows, may be seen eagerly feasting on the dry seeds which still remain on the withered plants. The soft grey down of the hoary leaves, later on in May and June, is used as linings for the nests of the Humming-birds, and other small birds that weave dainty soft cradles for the tiny families that need such tender care. Taught by unerring wisdom, each mother-bird seeks its most suitable material, and appropriates it for the use and comfort of its unknown, unseen brood. Let us not despise the common Mullein, for may it not remind us of Him who careth for the birds of the air, and giveth them from His abundant stores their meat in due season ; and that wonderful unerring wisdom that we call instinct : " Who *least*, hath *some* ; who *most*, hath *never all*." Happy are the wild flock for whose untold wants He provideth. The birds of the air teach us wisdom, for they obey the Creator's will, and rebel not at His laws.

## FALSE FOX-GLOVE.—*Gerardia quercifolia*, (Pursh).
### (PLATE V.)

I think old Gerarde, the first English writer on the wild flowers and native plants of England (for whose memory all botanists feel a sort of veneration) would have given a far better description of the stately plant honoured by his name, than the writer of this little work can hope to do, seeing that the only native species that has come within her knowledge, is a slender purple-flowered Gerardia, *G. purpurea* which grows on the margin of Rice Lake, among wild grasses and other herbage.

It has been said by one who was a diligent botanist and naturalist, (the late Dr. G. G. Bird) that no Gerardias were found north of the Great Lakes ; but all were confined to the Western and Eastern States ; this however was a mistake.     At that date very little was known of the Canadian Flora.

It was the trying time of pioneer life in the backwoods, when little heed was taken of the vegetable productions of the country, and even the trees of the forest were hardly distinguished by name, much less were the wild flowers cared for, unless some of the settlers knew of curative medicines to be extracted from the leaves or roots, or of some household dye for the home-spun flannel garments, which were then all that could be obtained as clothing for their families.     But to return to my Gerardias, several fine species have been found growing on the Islands of Lake Ontario, and on the banks of the Humber, that fruitful wilderness of many flowers ; and doubtless these handsome showy plants are well known in many localities westward in the Dominion of Canada.

The handsomest of all is *G. quercifolia*, Oak-leaved Gerardia, a robust, stately plant of from three to six feet in height, with large open-throated orange bells; it is known as False Fox-glove.   There are several fine purple-flowered species, and others of paler yellow than *quercifolia*, with stems coarse, rigid, downy or bristly ; the leaves mostly rough on the surface, and of a dull green.

I am not aware of any particularly useful qualities attributed to this Genus, but as ornaments to our gardens they would prove very attractive—one of the most suitable is *G. pedicularia*, a very much branched species which grows in dry thickets ; it is about 2 feet high, has prettily lobed foliage and a profusion of yellow flowers.   It seems a pity that these beautiful plants should be passed by only as weeds, unnoticed and unvalued.

GAY-FEATHER—BUTTON SNAKE-ROOT—*Liatris cylindracea*, (Michx.)

This pretty purple flower is found growing on dry hills, near lakes and rivers, on sandy flats and old dried water-courses.   The slender, stiff, upright stem is clothed with rigid, narrow, grass-like, dark green leaves, the longest being nearest to the root.   The flowers form a long spike of densely-flowered purple heads; the scales, of the involucre that surrounds them, are green tipped with black, and finely fringed ;

the styles protrude beyond the tips of the corolla. The root is a round corm, about the size of that of the Crocus, sweetish and slightly astringent, mealy when roasted, and not unpleasant to the taste. The roots are sought after by the Ground-hogs, which animals often make their burrows near the place where the plants abound, this is often on the slopes of dry, gravelly hills. At any rate it is on the sides of ravines, on the dry plains above Rice Lake, and on islands in our chain of back lakes in Burleigh and Smith, where I have found the bright Gay-feather blooming in the hot month of August. The seeds are hairy, almost bristly, of a light sandy brown when ripe. The blossoms retain their beautiful colour when quite dry, even for many years, and may be mixed with the flowers of the Pearly Everlasting for Winter bouquets or ornamental wreaths.

One of the species of this family, *L. scariosa*, a handsome flower found on our North-western prairies, is known by the name of Blazing Star. The showy flowers of the *Liatris* family, and their hardy habits, make them desirable plants for cultivation. They are easily propagated from seed.

Golden-rod—*Solidago latifolia,* (L.)

The Solidagos are among our late August and September wild flowers, coming in with the hot Summer suns, which have given the ripened grain to the cradle scythe of the harvest-man. The Trilliums and Lupines and gorgeous Orange Lilies have gone with the Moccasin-flowers, the sweet-scented Pyrolas, and the Wild Roses. Many of the fair flowers have faded and gone, but we are not quite deserted, we have yet our graceful Asters, our pretty Gay-feathers, our Sun-flowers, Cone-flowers and our blue Gentians, and brightening our way-sides with many a gay, golden sceptre-like branch, our hardy, sunny Golden-rods ; varying in colour from gorgeous orange to pale straw-colour , from the tall stemmed *S. gigantea* to the slender wand-like forms of the dwarf species, of which we possess many kinds, some with hoary foliage, others with narrow willow-like leaves of darker hue. On the grassy borders of inland forest streams we find the Golden-rods ; they seem to accommodate themselves to every kind of soil and situation. The rocky clefts of islands are gay with their bright colours, the moist shores of lakes, the sterile, dusty waysides, corners of rail-fences or the forest shades, no spot so rude but bears one or another species of these hardy plants. A coarse but grand Genus and not without its value. Not for ornament alone is the Golden-rod prized. The thrifty wives of the old Canadian settlers prized it as a dye-weed, and gathered the blossoms for the colouring matter that they extracted from them, with which they dyed their yarn yellow or green.

One of the late flowering species, *S. latifolia*, is remarkable for its fragrance, it is slender in habit, the lax branches trailing upon the ground in grassy woodlands. The leaves are large, very sharply and coarsely toothed, margined on the leaf-stalk, terminating in a slender point at the apex. The blossoms, which are larger than those of many of the taller species, are clustered in the axils of the large thin leaves at rather distant intervals along the slender branches; the silky pappus of the winged seeds is tinged with purplish-brown, the flowers are golden-yellow.

STRAWBERRY BLITE—INDIAN STRAWBERRY.—*Blitum capitatum*, (L.)

The Strawberry Blite—or, as it is often called, Indian Strawberry— is widely spread over the Northern States, and Canada.

Wherever the forest has been cleared it is sure to appear, as it seems to affect the rich black leaf-mould of the newly-cleared forest.

It is not indeed found within the close thick forest, but appears wherever a partial clearing has been made. It may be seen close to the rough log walls of the lumberer's or chopper's shanty, flourishing in great luxuriance under this half culture. On forest land, that has been burnt over and left uncropped, it may be seen in perfection; and within the garden enclosure, where it becomes a common weed : though truly more ornamental than many a flower that the gardener cultivates with care and trouble.

When fully ripe, the long spikes of crimson fruit, and foliage of a bright green colour, have a beautiful appearance, and tempt the hand to pluck the richly-coloured seed clusters ; but beauty is not always to be trusted, and in this case the eye is deceived and the taste disappointed The fruit is insipid and flavourless, though not unwholesome.

The red juice is used by the Indian-women in dyeing—and in old times the backwoods settlers made it a substitute for ink—but unless the colour be fixed by alum, it fades and disappears from the paper.

The Indian Strawberry or Blite, belongs to the Spinach family, and may be used with safety as a substitute for the garden vegetable, being perfectly harmless.

I well remember, many years ago, greatly alarming some of my neighbours in the backwoods, by gathering the tender leaves and shoots of these plants, and preparing them for the table. I was assured that death would be the result of my experiment; but I was confident in the innocent qualities of my fruit-bearing Spinach, and laughed at the drediction that I should find death in the pot.

Nor is the Indian Strawberry the only member of the Spinach tribe that is found growing in Canada. We possess several others, among these are the herbs commonly known by the country people as Good King Henry *(B. Bonus Henricus)*, which has been introduced from Europe, and Lamb's-Quarters *(Chenopodium album)*, which plants are still made use of as Spring vegetables, though not now in such repute as formerly. Happily few houses, or even shanties, but can boast of a garden around the dwelling. But many years ago it was a rare thing to see even a Cabbage-plot fenced in about the homestead, and the cultivation of flowers was regarded as a piece of useless extravagance, a mark of pride and idle vanity. We do not wish *those* good old times back again !

The leaves of the Indian Strawberry are thin, long-pointed, somewhat halbert-shaped, with shallow indentations at the edges. They are of a bright lively green colour. In the earlier stages of growth, the flowering spikes stand upright, but as the fruit ripens they decline, and are bending or entirely prostrate, much resembling the drooping Amaranth (called Love lies Bleeding) of our gardens, but more brilliant in hue. The berries of the Indian Strawberry are wrinkled on the surface and dotted over with purplish-black seeds, which lie embedded in the soft fruity pulp of the altered clayx in a manner similar to the Strawberry. The fruit begins to ripen in July, and continues by a succession of lateral branches to bear its red clusters all through August, and till the frosts of September cut it off and destroy the beauty of the plant.

TURTLE-HEAD—SNAKE-HEAD—*Chelone glabra*, (L.)

This coarse, but rather showy plant, is found in damp thickets near lakes and streams. The large, white, two-lipped flowers grow in terminal clusters or spikes ; the upper lip projects downward like a Turtle's bill ; the foliage is dark green, the leaves opposite, the edges coarsely-toothed, long and sharp-pointed ; the stem, simple, or widely branching and bushy ; the large handsome white flowers are often tinged with red or purplish-red ; the blossom is open-throated, somewhat contracted at the mouth by the overhanging of the upper lip. The whole plant is from two to three feet high. The name of the Genus is derived from a Greek word, which signifies a Tortoise, the form of the beaked corolla, resembling the head of a reptile, hence also the common name Snake-head, from the fancied likeness to the open mouth of a snake. The flowering season is from July to September ; probably, under cultivation, this flower would become highly ornamental as a large border-plant.

There are many very ornamental flowers belonging to the same Natural Order as the Turtle-head, among which are the Beard-tongue

(*Pentstemon*), Monkey-flower (*Mimulus*), Snap-dragon (*Antirrhinum*), Scarlet-cup (*Castilleia*), and Gerardia, with many other plants more remarkable for beauty than for any useful or healing qualities, but very showy in the garden, and not difficult of cultivation.

### CARDINAL FLOWER—*Lobelia cardinalis*, (L.)

One of the most striking of our native flowers is the Red Lobelia or Cardinal-flower. The plant had found its way into English gardens as a rarity before I saw it growing in all its wild beauty on the shores of the Otonabee, on my first journey, or rather voyage, up the country. There, growing at the edge of the low, grassy flat, beside the water—its tall, loose, spike of deep red flowers fluttering in the breeze and reddening the surface of the bright river with the reflection of its glorious colour—this splendid flower first met my admiring eyes.

It was but a short time before that I had seen it cultivated as a new and rare border flower, and here it was in all its loveliness on the banks of a lonely forest stream which then flowed through an almost unbroken wilderness, growing uncared for, unsought for and unvalued. The people—they were a rude set of Irish settlers—were amused at the delight with which I plucked the flowers. They cared for none of these things : they were, to them, only useless weeds.

There are several varieties of the Cardinal-flower, occasionally found among the wild plants near the inland lakes and creeks of the back-woods : some with flesh-coloured corollas, or white striped with red ; but these variations are not very common. The prettiest of the blue-flowered plants of the Lobelia family is a small, delicate, branching one with azure-blue and white petals, which is cultivated in hanging baskets, as its bright blue flowers and slender leaves droop gracefully over the pot or basket, and contrast charmingly with larger flowers of deeper colour and more vivid foliage.

The largest and most showy, but not often cultivated, of the Lobelias, is *L. syphilitica*, a stout-stemmed, many-flowered species, which is chiefly found near springs ; the flowers are full blue and the spike much crowded ; the height about eighteen or twenty inches ; leaves light green. The plant seems to flourish in clayey soil near water. Another blue-flowered Lobelia of slenderer habit is *L. spicata*, the leaves growing up the wand-like stem in threes, with intervals between ; and it has a one-sided look. The spike of flowers is loose and scattered, leaves very thin, long and narrow, light-green and smooth.

Though by no means so showy—for indeed it is a very simple looking flower—but more remarkable for its uses and medicinal qualities is the celebrated

PLATE IV.

I.   WILD LUPINE   (*Lupinus perennis*).

II.   MARSH MARIGOLD   (*Caltha palustris*).

### INDIAN TOBACCO—*Lobelia inflata*, (L.)

This plant is much sought after by the old settlers, and by the Indian medicine-men, who consider it to be possessed of rare virtues, infallible as a remedy in fevers, and nervous diseases. At first it has the effect of producing utter prostration of the nervous system, and is known to be of a poisonous nature. It is, I suppose, a case of "kill or cure."

A decoction of the dried plant relieves fever through the pores of the skin ; but though used by some of the old settlers, it should not be administered by any one inexperienced in its peculiar effects. The Indians smoke the dried leaves, from which fact the common name is derived--Indian Tobacco. They also call the plant Kinnikinik, which I suppose means good to smoke, as the word is also applied to one of the Cornels, and also to the aromatic Winter-green—the leaves of these plants being used as a substitute for the common Tobacco, or to increase its influence when smoking the "weed."

The Indian Tobacco is a small branching biennial, from nine to eighteen inches high ; leaves ovate-lanceolate, light green ; seed vessel inflated ; flowers pale blue, veined with delicate pencilled lines of a darker hue ; soil, mostly dry woods or open pastures ; nature of this innocent looking herb, a virulent poison.

### INDIAN PIPE—*Monotropa uniflora* (L.).

This singular plant has many names, such as Wood Snow-drop, Corpse-plant, and Indian Pipe. The plant is perfectly colourless from root to flower, of a pellucid texture and semi-transparent whiteness. There are no green leaves, but instead, broad and pointed scales, clasping the rather thick stem, which is terminated by one snowy-white flower. The flower, when first appearing, is turned to one side, and bent downwards, but becomes erect as it expands its silvery petals, these are five in number; stamens from eight to ten; stigma about five-rayed; seed vessel, an ovoid pod, with from eight to ten grooves ; seed small and numerous. Though so purely white when growing, the whole plant turns perfectly black when dried ; even a few minutes after they are gathered, as if shrinking from the pollution of the human hand, they rapidly lose their silvery whiteness and become unsightly. To see this curious flower in its perfection you must seek it in its forest haunts, under the shade of Beech and Maple woods, where the soil is black and rich ; and there, among decaying vegetables, grows this flower of snowy whiteness.

There are two species of the family. In a Hemlock wood I found the equally singular

G

PINE SAP—*Monotropa Hypopitys*, (L.)

A tawny-coloured, scaled, leafless species, with several flowers, covered with soft, pale yellowish-brown wool, fragrant, and full of honey, which fell from the flower cups in heavy, luscious drops. This plant is of rather rare occurrence, and only found here in Pine or Hemlock woods, though Gray speaks of it as common in Oak and Pine woods.

## GENTIANS.

> " And the blue Gentian flower that in the breeze
> Nods lonely ; of her beauteous race the last."—*Bryant.*

This interesting floral family takes its name from Gentius, a King of Illyria, who is said to have been the first to discover and be benefitted by its sanative properties.   The root used in medicine, is, I believe, a native of Spain.   The Alpine Gentian—so often spoken of by tourists—is of low stature, with very large, intensely-blue upright bells ; " a thing of beauty and a joy for ever," even to have beheld it growing in its serene loveliness on the edge of the icy glaciers and rude moraines of the Swiss Alps.

Of all our native flowers, the Gentians are among the most beautiful, from the delicately fringed azure-blue (Bryant's flower) to the fair, pale, softly-tinted, Five-flowered Gentian, with its narrow bells and light-green leaves.   All are lovely in colour and form, but none more deserving of our attention than the large-belled Soapwort Gentian, known also by the poetical name of

CALATHIAN VIOLET—*Gentiana Saponaria*, (L.),

This is the latest of all our wild flowers, it comes early in the Fall of the year, and lingers with us

> " Till fairer flowers are all decayed,
>     And thou appearest;
> Like joys that linger as they fade,
>     Whose last are dearest."

On sandy knolls, among fading grasses and withered herbage of our Oak-plains, we see the royal deep blue, open, bells of this lovely flower, its rich colour reminding one of a Queen's coronation robes.

This species somewhat resembles the European *G. Pneumonanthe,* (Linn.), which is also known by the same poetical English name.   In Sowerby's " English Botany," under the head of the last named species, we find :   " This pretty little plant is worthy of cultivation, and is quaintly mentioned by Gerarde, who says : ' the gallant flowres hereof bee in their bravery about the end of August,' and he tells us that ' the later physitions hold it to bee effectual against pestilent diseases, and the bitings and stingings of venomous beasts.'"

Our Gentians are the last tribute with which Nature decks the earth—her last brightest treasures—ere she drops her mantle of spotless snow upon its surface.

We find our latest flowering Gentian early in September, and as late as November, if the season be still an open one, it may be seen among the red leaves of the Huckleberry and Dwarf Willows, on our dry plains, above Rice Lake, and farther Northward. The Gentians seem to affect the soil on rocky islands and gravelly, open, prairie-like lands, among wild grasses. The finest, most luxuriant plants of *G. Andrewsii*, were gathered on islands in our back lakes, growing in rich mould in rocky crevices. The Five-flowered Gentian may be found on dry banks and open grassy wastes, while again the exquisite, azure-blue, single-flowered, Dwarf Fringed Gentian, *Gentiana detonsa* (Fries), prefers the moist banks of rivulets and springs. In drier places may be seen the stately, many-flowered, taller, blue Fringed Gentian, *G. crinita* (Frœlich.) There is also a charming intermediate form of *G. crinita*, about a foot high, with fewer flowers, but of a richer, fuller azure tint. It is of the Fringed Gentian that the poet, Bryant, writes :—

> Thou blossom bright with Autumn dew,
> And coloured with heaven's own blue,
> That openest when the quiet light
> Succeeds the keen and frosty night.
>
> Thou comest not when Violets lean
> Oe'r wandering brooks and springs unseen ;
> Thou waitest late, and comest alone
> When woods are bare and birds are flown,
> And frosts and shortening days portend
> The aged year is at an end.
>
> Then doth thy sweet and quiet eye
> Look through its fringes to the sky ;
> Blue, blue as if the sky let fall,
> A flower from its cerulean wall."—*Bryant.*

But bewildered among so many beauties, I have wandered away from my first love. The large dark-blue or open-belled Gentian *Gentiana Saponaria* (L). The leaves of this species are somewhat clasping at the base, and pointed at the end, at first green, but assuming a purplish-bronze hue ; the smooth stem is also of a reddish purple, with the large open five-cleft dark-blue corollas terminal on the summit, generally three blossoms, and between the axils of the leaves three

or more somewhat smaller bells may be found at intervals clustered on
the flower stem.      The beautifully-folded, deep purple buds are surrounded by the pointed bracts and leaves.

This species is less marked than *G. Andrewsii* (Griseb) by the
toothed appendages between the lobes of the flower; the absence of
these plaited folds gives our plant a wider, more open flower, which
renders it more attractive to the eye of the florist.      There is something
almost disappointing in the closed sac-like blossom of the

### CLOSED GENTIAN—*Gentiana Andrewsii,* (Griseb.)

Lovely as it is, one would like to peep within the closed lips, which
so provokingly conceal the interior.      The tips of the corolla are white,
but the sac-like flower is of a full azure-blue, striped in some cases with
a deeper colour.      There are often as many as five buds and blossoms
clustered at the summit of the flower stem, and in the axils of the deep
green, smooth and glossy leaves.

On parting the lips of the closed corolla we see at the narrowed neck
some toothed and sharply jagged appendages, which also may be observed
in many others of the Gentians, in greater or lesser degree.      This handsome
species is about eighteen inches high, with flowers more than an inch
in length, and loves rich leaf-mould near water on rocky islands.

### FRINGED GENTIAN—*Gentiana crinita* (Frœl).

Of the Fringed Gentians, we boast three forms, all charming and
attractive, and it seems strange that such beautiful flowers should not
have found their places long ere this in our gardens.      The seeds would
not be difficult to obtain from the tallest plant *G. crinita,* as it blooms
earlier and ripens its pods before the heat of the Summer has entirely
given place to frosts.

I have generally found the tall Fringed Gentian on dry, rather
gravelly soil, and river banks.      The buds of this flower are beautifully
folded, almost twisted, and are terminal, growing singly, on long foot
stalks; the corollas rarely unfold fully; the plaited folds are inconspicuous
or absent.      The colour of the flower of this tall species is light blue,
and white at the base; the upper edges of the corollas are elegantly
fringed and cut.      Though taller, and the bells more abundant, the lower,
deeper coloured fringed varieties are more lovely.

There is a bitter principle in the roots of most of the Gentians:
especially is it strongly developed in the Five-flowered Gentian
—*G. quinqueflora,* (Lam.)    This bitter principle is one of the character-

istics of the family, and probably our native plants might prove as valuable as tonics as the foreign root, were they tested. The Five-flowered Gentian is very unlike the bright and more showy blossomed species described above. The flowers in fives, are narrow bells of a delicate pale lilac-tint, clustered in the axils of the narrow, light-green leaves ; it is found sometimes on dry, grassy banks, and in the angles of fences by roadsides.

I have a specimen closely resembling the above species, sent from Iowa, the chief difference being that the tips of the slender flower-tubes are of a deep dark blue—our Canadian flower being only slightly tinted with very pale lilac. I have never found any of the Gentians growing in the forest, though several species seem to flourish in partial shade in open thickets.

With the Gentians I have brought to a close the floral season of the Canadian year. A few stragglers may yet be found amongst late Asters and Golden-rods, in sheltered glens and lonely hollows, but the glory of the year has departed : gone with the last deep blue bell of the loveliest of her race, the Calathian Violet, the solitary flower of the Indian Summer. All that now remains for us is the bright frosted foliage of the Dwarf Oaks and the scarlet tinged leaves of the low Huckleberry bushes; the brilliant berries of the leafless Winterberry, *Ilex verticillata* (Gray), and the clustered garlands of the climbing Bitter-sweet, *Celastrus scandens*, which hang among the branches of the silver-barked Birch and other forest trees, or near the margin of lake, or stream; and the crimson fruit of the frost-touched High-bush Cranberry *Viburnum Opulus*—while on dry, stony hills and rugged rocks the Bearberry covers with its creeping branches of dark green, shining, leaves and gay scarlet fruit, the scanty soil from which it springs. Let us prize them, for from henceforth till the tardy Spring revisits the earth, its treasures of leaf and blossom will be to us as a sealed book bound up in ice and snow. No more are we tempted by verdant wreaths of glossy leaves or gaily tinted-flowers. We must be contented with wintry landscapes, snow-flakes and frost-flowers, and the crystal casing that covers the slender branches of the Birches and Beeches, or hangs in diamond drops on the tassels of the Spruces and Balsam Firs.

Tread softly, traveller, lest the transient glory of our Frost-flowers dissolve at your feet. Emblems of earthly beauty, earthly riches and earthly fame. But there are brighter gems and fairer flowers of heavenly growth that fade not away, but which will flourish in the Paradise of God more glorious than the fairest beauties of our earthly home.

## GRASSES.

" And God said let the earth bring forth grass,
      The herb yielding seed."
\*      \*      \*      \*      \*      \*      \*      \*      \*      \*
" And the earth brought forth grass,
      And herb yielding seed."—*Gen. I,11-12.*

In drawing this little volume on the native plants to a conclusion, though many have been left unnoticed or unknown by me, I must say a few words respecting the Grasses. Not indeed to add a botanical description of this most beautiful and graceful tribe of plants, which deserves a volume from the pen of one who has given great attention to the subject, and which seems to me to require the knowledge of a scientific botanist. To do justice to that I must confess I am not competent ; any knowledge that I possess is simply that of an observer and a lover of the beautiful works of my Creator.

The student of botary will not be content merely with my superficial desultory way of acquiring a more intimate acquaintance with the productions of the forest and the field ; and to such I would recommend a more particular study of our beautiful native Wild Grasses, including the Rushes, and the Sedges. At present the field has not been entered upon fully, if even its very borders have been gleaned, unless by that industrious and indefatigable botanist, Professor John Macoun, whom we might well call the Father of Canadian Botany.

But though I cannot venture to treat the subject of the Grasses as a botanist, I cannot pass them by, without introducing a few of the lovely graceful things to the notice of my readers. And if my remarks should prove rather desultory in their range from Prairie to Forest, and from Field to Lake, or from swampy bank of Creek or Marsh, I beg my friends to bear with me a little while.

Drooping gracefully in wide branching panicles, we find on our wild plains a soft pale-flowered grass, known by the Indians as Deer-grass, *Sorghum nutans,* (Gray.) in the herbage of which the Deer found (for it is a thing of the past) both food and shelter. The husk or glumes of this beautiful grass are hairy or minutely silky, which gives a peculiar soft greyish tint to the bending pedicels of the pale spikelets. The culm is from three to four feet high, the leaves hairy at the margins. Another grass, *Andropogon furcatus* (Muhl.) more showy but not so graceful, being more upright in its habit of growth, differs very much from the above. This grass is tall, jointed, stiffer in the stem, leaves of a brighter green, heads of flowers spiked, but also branching ; glumes of a rich red-brown, made more conspicuous by the bright golden yellow anthers. This grass is also a Plain grass, and known by the same familiar name as the former ; the Indians say "Yes, both Deer-Grass ; Deer like that too." It was to increase the growth of this grass that the Indians, at intervals of time, set fire to the Rice Lake Plains on the high plateau of land to the eastward, where there was a great feeding ground for the Deer and their fawns. For many years this tract of land was covered with Oak-brush, with only a few old trees that had escaped being injured by the fire. Now, indeed, we have noble Oaks of many species, fine branching, well developed trees of White, Black, Red, Scarlet, and Over-cup Oaks, that adorn the Plains and form avenues of the concession and side-lines, most ornamental and grateful to the eye of the traveller. It must have been nearly a century ago since these Plains were last burnt over—not within the memory of the oldest settler in the Township of Hamilton. Yet deep down, some six or seven feet below the surface, the charred remains of Oaks are found to prove the truth of the Indian name, " The Lake of the Burning Plains." Indian names have always some foundation ; adopted from peculiar circumstances, they have acquired a sort of historical value among the people.

The name of " Rice Lake " is derived from the fields of Wild Rice *Zizania aquatica*, (L.) which abound in the shallower waters of this fine inland sheet of water, and give the appearance of low verdant islands clothing its waters. When the Rice is ripened, and the leaves faded, a golden tint comes over the aquatic field, and the low Rice islands as they catch the rays of the sun take the form of sands glowing with yellow light. Where the water is low, these Rice beds increase so as nearly to fill the shallow lakes and impede the progress of boats, changing the channel and altering the aspect of the waters.

In the month of June the tender green spikes of the leaves begin to appear ; in July the Rice begins to push up its stiff, upright stalk ; sheathed within its folds are the delicate, fragile flowers ; from the slender glumes, the beautiful straw-coloured and purple anthers hang down, fluttering in the breeze which stirs the grassy leaves that float loosely upon the surface of the water, rising and falling with every movement. The plant grows in lakes, ponds, and other waters, where the current is not very strong, to the depth of from three to eight feet or even deeper. The grassy or ribband-like flexible leaves are very long. I remember a gentleman who was rowing me across the lake drew up one at a chance on his oar and measured it, the length being eleven feet ; but with the culm and flower it would have measured twelve or thirteen feet in length.

The month of September or later, in October, is the Indian's Rice harvest. The grain, which is long and narrow and of an olive green or brown tinge, is then ripe.    The Indian-woman (they do not like to be called squaws since they have become Christians) pushes her light bark canoe or skiff to the edge of the Rice-beds armed not with a sickle, but with a more primitive instrument—a short, thin-bladed, somewhat curved, wooden paddle, with which she strikes the heads of ripe grain over a stick which she holds in her other hand, directing the stroke so as to let the grain fall to the bottom of the canoe ; and thus the Wild Rice crop is reaped, to give pleasant, nourishing and satisfying food to her hungry family.

There are many ways of preparing dishes of Indian Rice : as an ingredient for savoury soups or stews ; or with milk, sugar and spices, as puddings ; but the most important thing to be observed in cooking the article is steeping the grain—pouring off the water it is steeped in and the first water it is boiled in, which removes any weedy taste from it. It used to be a favourite dish at many tables, but it is more difficult to obtain now.

The grain all collected, it is winnowed in wide baskets from the chaff and weedy-matter, parched by a certain process peculiar to the Indians, and stored in mats or rough boxes made from the bark of the Birch tree—the Indian's own tree.    Formerly we could buy the Indian Rice in any of the grocery stores at 7s. 6d. per bushel, but it is much more costly now, as the Indians find it more difficult to obtain.    Confined to their villages, they have no longer the resources that formerly helped to maintain them.    The birch-bark canoe is now a thing of the past ; the Wild Rice is now only a luxury in their houses ; by and by the Indians also will disappear from their log-houses and villages and be known only as a people that were, but are not.    I am not aware of any other edible grain that is indigenous to Canada.    The Fox-tail, *Setaria viridis*, (Beauv.) indeed, has hard seeds, but it is utilized only in some places where it abounds (to the farmer's great disgust) as food for his hogs and fowls. The marsh-growing Red-top or Herd-grass, *Agrostis vulgaris*, (With.) is used as hay.    We have many other wild, coarse grasses also that are harvested ; and the prairies abound with nutritious plants of this Order which are a great resource for the support of the cattle during all seasons. What would become of the settler's beasts in the North-West Provinces but for the Prairie Hay ?    Very beautiful varieties of the lovely Prairie-grasses have been gathered by kind friends and sent to me from this " Wild North Land."

One, the cruel Arrow Grass, *Stipa spartea*, (Trin.) is a great nuisance to the settler, the barbed shafts and curiously twisted stipes piercing hands and feet or insinuating their hard points into the flesh or clothing.    The long, twisted arrows of this grass have a curious fashion of winding themselves together, forming a sort of hard rope ; the barbed seed lies below, attached to these twisted arrows.   There is also on the prairies a wild grass known by the descriptive name of Porcupine Grass ; possibly the Arrow Grass may be the same plant with another name.   But turning from this uninviting Prairie Pest, as the settlers call it, I would rather call attention to the useful and sweet-scented Indian Grass, which supplies the poor Indian-woman with the material which she weaves into such lovely, tasteful, ornamental baskets, now almost her only resource for materials for her basket-work, by which industry she can earn a small addition to her scanty means of obtaining food and clothing.   Were it not going beyond the bounds of my subject I might plead earnestly in behalf of my destitute, and too much neglected, Indian sisters and dwell upon their wants and trials ; but this theme would lead me too far away from my subject.  The Indian Grass, so called *Hierochloa borealis*, (Roem. & Sch.) is little known in its native state, as it is only the Indians themselves who know where to seek for it.   This is among lonely lakes and forest haunts. The soil where it grows is low, sandy flats, especially on shores where the soil is composed of disintegrated friable rocks, reduced to gritty, coarse sand, where it can send up its slender, white, running roots most freely ; and there it sends up early in May its culms and light panicles of shining flowers ; the glossy straw-coloured  plumes and purple anthers make this grass a very lovely object.   The leaves, too, are of a shining bright full green.   It is the earliest of any of the grasses to push up its pointed blades above the ground ; and, as far as my knowledge of the plant goes, for I have had it in my garden for many many years, it is the earliest to blossom.   Only when dried, or rather withered, does it give out its sweet scent, which it retains for years.

I have braided the long ribband-like leaves and made  dinner-mats of them, and also chains tied with coloured ribbon, after the Indian fashion and sent them to friends in the Old Country to lay like Lavender in their drawers.   One thing I must observe of the Indian Sweet-grass, although it grows readily, and flourishes in any odd corner of the garden in which you plant it, it rarely puts forth a flowering stem, nor can I account for this unless it may be the absence of some speciality in the native soil that is lacking, and for the need of which it may grow luxuriantly as to leaf but brings no fruit to perfection.

Among the common wild grasses we have many  kinds  known by such expressive names as Red-top, Blue-joint, Herds-grass, Beaver Meadow-grass, Wild Oats, Wild Barley,  Fox-tail, Squirrel-tail, Poverty-grass, Cock's-foot, Couch or Spear-grass, Millet, with many others, named or unnamed, that are peculiar to certain localities, in open fields, in the shade of the forest, the thicket, the banks of creeks, in water, or on dry waste lands ; there is no spot but has some Grass, or Rush, or Sedge, or Reed ; they spring up by the water-courses, on the dry parched sands of desert places, and in our path by the way-side ; thus we find this lowly herb, under some distinguishing form, wherever we go. Is it not intended as a silent monitor to remind us of the frailty of our earthly being, by bringing back to us the words of the Psalmist : " As for man his days are as grass, as a flower of the field so he flourisheth, for the wind passeth over it and it is gone, and the place thereof shall know it no more."—Ps. 103.

How often in the inspired words do we find similar allusions made to the grass in language alike practical and touching.

" The voice said Cry ! And he said What shall I cry ? "

" All flesh is grass and all the goodliness thereof is as the flower of the field."

" The grass withereth, the flower fadeth, * * * * but the word of our God shall stand for ever."—Isaiah XL, 6–7.

Thus the grass that we tread beneath our feet, as well as the fairest flower, has alike a significance and a teaching to lead us up to the throne of Him who makes the grandeur of the heavens above, and the lowliest plant on earth, to speak to us of His goodness, His wisdom and His fatherly care for all. Let me close with the lesson of faith that Christ the Lord himself gave to his disciples : " If God so clothe the grass of the field, * * * * shall He not much more clothe you, O ye of little faith ?"

# A FAMILIAR DESCRIPTION

## —OF THE—

# Flowering Shrubs of Central Canada.

―――

"Hie to haunts right seldom seen,
Lovely, lonesome, cool and green.
Hie away, hie away,
Over bank, over brae,
Hie away."—*Waverley.*

LEATHERWOOD—MOOSEWOOD—*Dirca palustris,* (L.)

THE Leatherwood or Moosewood is one of the very earliest of our native shrubs to blossom; little clusters of yellow, funnel-shaped flowers appear on the naked, smooth-barked branches early in April; three or more buds project from an involucre of as many scales covered thickly with soft, brown, downy hairs. The leaves, which expand soon after the falling off of the flowers, are smooth, of a bright light green, oblong, entire, and placed alternately along the stems. This pretty, shrubby bush seldom exceeds five feet in height, but is often much lower. The bark is of a pale greenish-grey, very tough, and while fresh and young not easily broken; it becomes more brittle when thoroughly dried, losing its useful pliant qualities. The bush settlers used the tough bark in its green state as a substitute for cordage in tying sacks and for similar purposes. This hardy shrub is, I believe, the only native representative in Canada of the Mezereum family; it has neither the fragrance nor the dark glossy foliage of the Daphne or Spurge Laurel of the English gardens; but, nevertheless, forms a pretty addition to our garden shrubberies; the early blossom, abundant foliage, and light scarlet globular berries are very attractive. The New England people call the plant Moosewood in allusion to the hairy covering of the flower-buds. The Canadian's Leatherwood, and the Indian's *Wycopy* meaning a thong, on account of its tough leathery bark. The specific name, *palustris,* would imply that it was more particularly a marsh-loving plant; but the Leatherwood may be found frequently growing on dry gravelly ground, and is by no means confined to wet, marshy soil. Dr. Gray says : " The name of a fountain near Thebes was applied by Linnæus to this North American Genus for no imaginable reason, unless because the bush frequently grows near mountain rivulets.

This shrub is found all over the Eastern and Western parts of the Dominion and has a wide northerly range. I know of no especial uses excepting the one already named among the settlers in the back-woods and the Indians, who use the bark as loose handles for their bark baskets used in rough work.

### FEVER-BUSH—SPICE-BUSH—*Lindera Benzoin*, (Meisner).

This highly fragrant shrub is commonly found growing in low, wet, marshy ground, and is sought for by the Indians for medicinal uses ; the bark and twigs (for it is in them the aroma is contained) form one of their luxuries, mingled with tobacco. The spicy, sweet-scented wood long retains its flavour, even when dried, and is most agreeable. The bush is about four or five feet high ; the bark of the older branches grey and smooth, but the young twigs and leaf-stalks are blackish. The flowers in this, as in Leatherwood, appear in umbel-like clusters in April before the foliage is developed ; the blossoms are yellow, or honey coloured ; the leaves entire, very smooth, darkish green, oblong and pale underneath. This shrub belongs to the Laurel tribe, and is nearly allied to the Sassafras. The natives make a fever-drink of the twigs, besides chewing and smoking the bark.

### TRAILING ARBUTUS—MAY-FLOWER—*Epigæa repens*, (L.)

#### (PLATE II.)

The fragrant, graceful *Epigæa repens*, the sweet May-flower of the Northern States, and of our own Canada, is too lovely to be forgotten in these short floral biographies ; indeed, this pretty trailing ever-green is well deserving of a place amongst the most cherished treasures of the conservatory, for few exceed it in beauty, and none in fragrance. It is to be found within the Pine forests, beneath trees where but a scanty herbage flourishes ; and on dry, sandy and rocky ground we see its ever-green, shining, ovate leaves, and delicate pink flowers, covering the ground during the month of May. The Americans know it by the name of May-flower, so called from its season of blossoming ; in England it is a favourite green-house shrub under the name of Trailing Arbutus. The leaves rise on long foot-stalks from the somewhat horizontal branches, they are unequal in size, the largest being nearest to the summit ; the leaf-stalks are clothed with clammy reddish-coloured hairs, which contain an odorous gum ; the flowers are tubular, divided into five segments at the margin, in colour varying from white to rosy-pink ; the inside of the long tube is beset with silvery hairs. The lovely, waxy flowers are clustered at the summits of the creeping stems, and give out a delightful aromatic scent. The classical name of our pretty ever-green is derived from the Greek, and signifies—upon the earth—in allusion to its prostrate trailing habit.

### BEAKED HAZEL-NUT.—*Corylus rostrata*, (Ait.)

The Beaked Hazel-nut is a small bush, not more than three to four feet high ; the leaves are large, oval, and coarse in texture, furrowed and dentate, at the edge. The catkins appear in April ; the light crimson tufted pistillate flowers in May. The nut is enveloped in a rough green involucral calyx, which is undivided and closely invests it, this rapidly diminishes in size above the nut, and is prolonged for about an inch ; in shape it takes the form of a hawk's bill, whence the specific name *rostrata*, or beaked, is derived.

The calyx is closely beset with short, bristly hairs which pierce the fingers, producing an unpleasant irritation ; especially is this felt when the fruit is ripe, and the enveloping case is withered and dry. The nut is sweet and well-flavoured, and resembles the common Filbert more than the Wild Hazel-nut of England. The bush seems to affect dry open ground and copse woods. There is another native species, the

### AMERICAN HAZEL-NUT.—*Corylus Americana*, (Walt.)

This is a much taller bush, found chiefly in damp thickets ; the long, slender wand-like nut-brown branches springing from a thickened root-stock or stool, and reaching to a height of ten to fifteen feet in damp localities. The sweet nut is round and thick shelled, the involucral calyx spreading at the tips and more open than in the former species. The foliage is round, somewhat cordate, or heart-shaped, coarsely pointed and serrated. The flowers, which are of two kinds in this genus, come successively before the unfolding of the leaves. The two species are very distinct in their appearance and character. The Beaked Hazel-nut bearing more likeness to the Filbert, while the present species resembles the common Hazel-nut.

The classical name *Corylus* is derived from a Greek word, signifying a helmet, from the shape of the calyx.

### RED-BERRIED ELDER.—*Sambucus pubens*, (Michx.)

The red-fruited Elder is often confounded by ignorant persons with the *Rhus Toxicodendron*, to which the names of Poison Elder, Poison Oak, and Poison Ivy have been given, thus transferring the evil qualities of the poisonous *Rhus* to a perfectly harmless shrubby tree, which deserves to be redeemed from such slanders. The Red-berried Elder is widely distributed over the Dominion of Canada.

In every waste place; on old neglected fallows which have been subjected to the ravages of fire; in corners of fences, and even in gardens, if care be not taken to ruthlessly root out the intruder, this hardy native may be found. The panicles of greenish-white flowers may be seen in the month of May, among black and burnt stumps, and girdled Pines, enlivening the coarse verdure of the dull-green, pinnated leaves, and grey warty branches; the flowers of this species, as well as those of the Black-berried Elder, *S. Canadensis*, (L.) emit a faint but sickly odour. The flowers of the latter species are whiter, borne in much larger and flatter cymes, and do not appear until June.

The embryo blossoms of the Red Elder are formed soon after the fall of the leaf in October, and may be distinctly seen in the large globular buds which adorn the bare branches in Winter; they are closely packed within the protecting cases, like hard-green seeds, each flower-bud perfect as if ready to unfold in the first warm sunshine; but not so, for the embryo flower must lie dormant in its cradle till the next Spring, when the warmth of the May sunshine opens it out to life and light. The blossoms are succeeded by an abundance of small berries, which, during the month of June, ripen, and adorn the landscape with their brilliant scarlet hues. The juice of the ripe fruit is a thin acid, slightly partaking of the peculiar flavour of the wood, not agreeable, but perfectly wholesome. The gay berries are a favourite food with wild birds, which soon strip the trees of their ornamental clusters.

### TWIN-FLOWERED HONEY-SUCKLE.—*Lonicera ciliata*, (Muhl.)

Though we have not, in Canada, the sweet-scented and graceful Woodbine of the bowery English lanes and hedge-rows—the theme of many a poet's lay—from Shakespeare and Milton down to Bloomfield and Clare—yet we have some charming flowering shrubs that are too lovely to be disregarded by the lover of Nature. Among our wild native species, there is not one more elegant than the Twin-flowered Honey-suckle, or Bush Honey-suckle. It is one of the earliest of our shrubs to unfold its tender light-green leaves. A few warm days in April—if the season be mild—and we may perceive the slender sprays assuming a welcome tint of verdure—the glad promise of Spring.

The ovate leaves, of pale green, are delicately fringed with silken hairs, at first of a slight purplish tinge. The flowers appear in pairs, connected twin-like from the axils of the leaves; in colour, something between a pale primrose and greenish-white, often tinged with purple. The elegant drooping bells are divided at the edge of the corolla, into five pointed segments, slightly turned outward, showing five stamens, and one style, which projects a little beyond the funnel-shaped flower. These graceful flowers united at the ovary, hang beneath the leaves on slender thready pedicels—so slight that the least breath of air swings their light fairy bells. One might almost be tempted to listen for some sweet music to issue from their hollow tubes. The twin berries, when ripe, are of a semi-transparent ruby-red, but like the fruit of all the Genus, they are tasteless or of a sickly sweet flavour. They form a feast for birds and numerous species of flies, which feed upon the pulp and juice. The country people give the name of "Fly Honey-suckle" to this shrub —doubtless from having noticed how attractive the fruit is to the insect tribes.

The Bush Honey-suckle thrives well in the garden under a moderate degree of shade, and in black vegetable mould.*

The general habit of this shrubby Honey-suckle is upright, not climbing; the branchlets are slender, with a pale greyish-green bark, and bend outwards, which gives a light and graceful aspect to the bush. The crimson, juicy berries are oblong, united at the base, and contain several yellowish, bony seeds.

### Small-flowered Honey-suckle.—*Lonicera parviflora*, (Lam.)

This pretty clustered trumpet Honey-suckle is also a native of our Canadian woods: a climber, but not often ascending to any great height, sometimes low and bush-like. It might be termed a dwarf climbing Honey-suckle. The flowers are showy and clustered in loose terminal heads; the tube very slender, and the segments of the corolla narrowly pointed.

This shrub seems to accommodate itself to circumstances, as it does not attempt to climb when transplanted to open ground, but forms a compact bush.

The abundance of its pale red and yellow flowers in light, graceful, clusters, and bluish-green foliage, make it a pretty ornament to the garden, to which it takes kindly when transplanted; the only dis-

---

* It is claimed to be a valuable remedy in cases of Dropsy.

advantages are the evanescence of its blossoms and its brief flowering season. The berries, however, are abundant, and are of a pretty light reddish-orange colour.

HAIRY YELLOW-FLOWERED HONEY-SUCKLE.—*Lonicera hirsuta,* (Eaton.)

This is a large, robust species ; the leaves large. ovate, and downy underneath ; the upper pair perfoliate, forming a boat-shaped involucre to the large, hairy, honey-coloured clusters of flowers, which are terminal. The stem of this rather handsome but coarse species is woody, branching and slightly twining ; the hairy, yellow trumpet-shaped flowers exude a clammy, sweet dew, which attracts numbers of flies which hover about them, with those honey-loving vagrants the Humming-birds. This species is chiefly found in open copses and on rocky islands. There are several other native Honey-suckles. Closely allied to the *Loniceras* is a pretty flowering shrub known as

FALSE HONEY-SUCKLE— *Diervilla trifida,* (Mœnch).

This shrub is often found on upturned roots in the forest, but it also flourishes in more airy situations, as the edge of open, cleared ground in the corners of rail fences, where it has access to sun-light and freer air. It seldom grows higher than two or three feet, forming a low leafy bush ; the leaves oblong, slightly toothed, in opposite pairs ; the branches are covered with a smooth, red bark ; the foot-stalks of the leaves are also red ; the flowers funnel-shaped ; the slender corolla divided into five lobes, the lower lip trifid. The flowers on slender peduncles, mostly in threes, spring from the axils of the leaves. The small seeds are contained in a hard two-celled, two-valved woody pod. The colour of the flowers varies from straw-colour to tawny yellow. Under cultivation the *Diervilla* increases in size and abundance of the flowers ; it is very hardy and will thrive in sunnier spots than the more delicate Twin-flowered Honeysuckle, which requires shade.

SNOW-BERRY.—*Symphoricarpus racemosus,* (Michx.)

Everyone is familiar with that pretty, ornamental garden shrub, the Snow-berry, so often seen in English shrubberies, as well as in our Canadian gardens ; but every admirer of it does not know that it is a native of the Dominion and may be found growing in uncultivated luxuriance on the banks of streams and inland waters, on the rocky banks of rapid rivers and lonely lakes, whose surface has never been ruffled by the keel of the white man's boat, spots known only to the Indian hunter or the adventurous fur-trapper. There, bending its flexile branches to kiss the surface of the still waters, its pure white waxen berries may be seen, looking as if some cunning hand had moulded them from virgin wax and hung them among the dark green foliage for very sport.

The blossoms of the Snow-berry are small, red and white bells, in clustered loose heads along the ends of the light, flexible sprays; during the flowering season the branches are upright but droop downward in Autumn from the weight of the large round snow-white berries. The brown, bony seeds lie embedded in the granular cellular pulp. Though quite innoxious, the fruit is insipid and more useful for ornament than for any other purpose, as far as man is concerned, but forms a bountiful supply of food to many of the birds that remain with us late in the Autumn. The plant multiplies by suckers from the roots and by seeds. The leaves are small, oval, slightly toothed, of a dull, dark bluish-green. This shrub is a native of all the Northern States of America, extending northward and westward in Canada. It belongs to the same Natural Order as the Honey-suckle, that lovely creeping plant the Twin-flower, and the Elders.

### SWEET-FERN.—*Comptonia asplenifolia*, (Ait.)

The popular name by which this shrub is known among Canadians —Sweet-Fern—is improperly applied and leads to the erroneous impression that the plant is a species of Fern. It is a member of the Sweet-Gale family and belongs to the Natural Order *Myricaceæ*.

The Sweet-Fern grows chiefly on light loam or sandy soil, in open dry uplands, and on wastes by road-sides, forming low thickets of small, weak, straggling bushes, which give out a delicious aromatic scent— somewhat like the flavour of freshly grated nutmegs—but the smell is evanescent, and soon evaporates when the leaves have been gathered for any length of time. The twig-like branches are of a fine reddish colour; the leaves are long, very narrow, and deeply indented in alternate rounded notches, resembling some of the Aspleniums in outline, whence the specific name. The flowers are of two kinds: the sterile in cylindrical catkins, with scale-like bracts, and the fertile in bur-like heads.

### SWEET-GALE.—*Myrica Gale*, (L.)

This sweet-scented low shrub may be found bordering the rocky shores of our inland Northern lakes in great abundance, and may be readily recognized by its bluish dull green leaves, and the fine scent of the plant. The leaves when stirred or crushed giving out a fine aroma of higher flavour, but resembling that of the Sweet-Fern, *Comptonia asplenifolia.* The sterile catkins, closely clustered, appear

H

before the leaves; the seed is contained in rough scaly heads; the leaves are toothed at the edges, broader at the upper end and narrowing at the base. The whole bush scarcely exceeds four feet in height, but throws out many small branches, and forms a close hedge like thicket near the margins of lakes and ponds; those lonely inland waters, where, undisturbed for ages, it has flourished and sent forth its sweetness on the desert air—"Just for itself and God." Yet the qualities of this shrub have not been quite overlooked by the native Indians, and by some of the old inhabitants of the back country, who use the leaves in some of their home-made diet drinks and in infusions for purifying the blood. As the luxuries of civilization creep in among the settlers, they abandon the uses of many of the medicinal herbs that formerly supplied the place of drugs from stores.

The old Simplers and Herbalists are a race now nearly extinct. I am inclined to agree with a statement I once heard, to the effect that hot stoves and doctors' drugs have fostered or introduced many of the diseases that carry our young people to an early grave, and have rendered the old ones prematurely infirm.

NEW JERSEY TEA—RED-ROOT.—*Ceanothus Americanus*, (L.)

There is an historical interest attached to the name of this very attractive shrub which still lingers in the memories of the descendants of the U. E. Loyalists in Canada and in the State of New Jersey, where the leaves of the *Ceanothus* were first adopted as a substitute for the Chinese Tea-plant. Even to this day Americans will cross to Ontario in Summer, to gather quantities of the leaves to carry back from our plains, where it is found in great abundance. And while they commend the virtues of the plant, they no doubt recount the tales of war, trouble and privation, endured in the old struggle waged by their grandfathers and great-grandfathers for independence, when, casting away the more costly tea, they had recourse to a humble native shrub to supply a luxury that was even then felt as a want and a necessity in their homes.

The leaf of the New Jersey Tea resembles that of the Chinese very much, and if it wants the peculiarly fragrant flavour that we prize so highly in the genuine article, yet it is perfectly wholesome, and if prepared by heat in a similar way might approach more nearly to the qualities of the foreign article. Indeed we are not sure but that it really does form one of the many adulterations that are mixed up with the teas of commerce, for which we are content to pay so highly. Many years ago I was applied to by persons in Liverpool to supply their firm with large quantities of the leaves, no doubt it was for the purpose of adulterating the foreign teas in which they dealt. Of course the proposal was declined.

An old friend, one of the sons of a U. E. Loyalist, told me that for some years after leaving the United States (the family were from Vermont) that the genuine Chinese Tea was rarely to be met with in the houses of the settlers, especially with such as lived in lonely backwoods settlements, that for the most part they made use of infusions of the leaves of the Red-root, or New Jersey Tea, as they had learned to call it, of Labrador Tea, *Ledum latifolium*, Sweet-Fern, *Comptonia asplenifolia*, Mountain Mint or other aromatic herbs, or even of the sprigs of the Hemlock Spruce.   Many of the old folks still retain a liking for the teas made from the wild herbs, and use them as diet-drinks in the Spring of the year with great benefit to their healths.

The light feathery clusters of minute white flowers of the *Ceanothus* have a charming appearance among the dark green foliage, and adorn the hills and valleys of the grassy Canadian plain-lands.   Where the soil is light loam.the shrubs are lower, and the flowers somewhat smaller, but very abundant, and give out a faint sweet odour.   In damper, more shaded spots, the flower-clusters are larger and borne on long foot-stalks. The leaves of the shrub are ovate, oblong, ribbed, and toothed at the edges.   The root is of a deep red colour, astringent and used medicinally.

The flavour of the leaves is slightly bitter and aromatic.   I consider this pretty *Ceanothus* to be one of the most ornamental of our native flowering shrubs, and well worthy of introduction into our gardens. Abundant clusters of delicate white flowers, that cover the bush during the months of July and August, have the appearance of the froth of new milk at a little distance.   The flowers are slender, the petals hooded, spreading, on slender claws longer than the calyx, which is five-lobed, coloured like the petals.   The seed-vessel is three-lobed, splitting into three parts when dry ; the seed is round, hard and berry-like. The branches and woolly stems wither and die down in Autumn, to be replaced by new shoots in the ensuing Spring.   In height the shrub varies from two to five feet.

WILD SMOOTH GOOSEBERRY—*Ribes oxyacanthoides* (L).

Our woods and swamps abound with varieties of the widely diffused Gooseberry and Currant family, and though at present neglected and despised, they no doubt could, by proper treatment, be made valuable and serviceable to man.   Of the Wild Gooseberries, there are several kinds.   The best and most palatable, being the smooth skinned, small purple Gooseberry, *Ribes oxyacanthoides* ; this is the least thorny of the Genus, and by cultivation, can be rendered a nice and serviceable fruit for preserving and other table uses.

It grows in low ground or on the borders of beaver meadows and damp thickets, and seems to be found in every part of the Dominion. The bush is low, not more than three to four feet, or less, with not very prickly stems, and smooth berries, generally in pairs ; the calyx of the flower purplish, and fruit when ripe of a dark purple colour ; leaves, smooth and shining, and pale beneath.

THORNBERRY—PRICKLY GOOSEBERRY—*Ribes Cynosbati*, (L.)

The fruit of this Wild Gooseberry is perfectly rough and spiny, and troublesome to gather, but in old times, was sought for by the settlers in the backwoods as a welcome addition to their scanty fare.   By scalding and rubbing the berries in a coarse cloth, much of the roughness was removed ; in its green state the berries were used in the form of pies and puddings, or, when softened, mixed with sugar and milk.   When ripe, it was made into preserves, but the harshness of the bristly skin was not very easily overcome, especially if the fruit was over-ripe.   Still it was one of the cheap luxuries that found a welcome place at the shanty table.   This is a tall bush from 4 to 6 feet in height, which grows in dry rocky woods, and bears a profusion of greenish bells, from one to three on each slender pedicel, in the month of May.

Another of our native Gooseberries is not so wholesome nor so useful ; this is the

SMALL SWAMP GOOSEBERRY.—*Ribes lacustre*, (Poir.)

Very pretty in flower, but very bristly, and the fruit small, not larger than peas, in slender racemes, of a pale red-colour, and unpleasant flavour.   The blossoms are pink and hang in graceful bunches on the weak and very prickly branches.   This small bristly species resembles the

TRAILING HAIRY CURRANT.—*Ribes prostratum*, (L'Her.)

This is the least desirable of the Currant family—being far from wholesome.   The whole plant is weak and reclining on the ground often rooting from the joints.   The leaves are rather large, smooth and 5 to 7 lobed.   The small, round, very pale red berries are hairy, glandular, and of a very unpleasant taste and odour.   I have known persons made very ill by eating tarts made of the Hairy Currants. It is easily distinguished by its trailing habit and hairy berries, and erect racemes of flowers.   I have found it chiefly growing in low lands and thickets, near swamps.   A larger bush and of common occurence, in swampy ground, is the

WILD BLACK CURRANT.—*R. floridum*, (L.)

When in blossom this Wild Black Currant is an ornamental object, The flowers of a pale greenish-yellow, are larger than the common garden species, and droop in long, graceful flowery racemes from the branches. The leaves are of a greyish-green, sharply lobed ; the bark grey and smooth ; berries very dark red, deepening when ripe to blackish-purple ; they are large and somewhat pear-shaped, in flavour not unlike the garden fruit. I should think it possessed of a narcotic quality ; certainly it is not very agreeable, though some people like it, and it is extensively used as a preserve. The bush takes kindly to cultivation but is, I think, more ornamental than useful.

WILD RED CURRANT.—*Ribes rubrum*, (L.)

Is said to be identical with our cultivated Garden Currant. In its wild state the fruit is small, very acid, and not unpalatable or unwholesome, but has a flavour of the astringent bark. This woody taste is common to many of our fruits in their natural state, but seems to be much reduced by care and cultivation.

JUNE-BERRY—SHAD-BUSH.—*Amelanchier Canadensis*, (T. & G.)

The June-berries are not only very ornamental shrubs but their fruit is very pleasant and wholesome, especially when mixed with acid berries, such as Currants and Cherries. The tallest of the Genus is the Shad-bush, which is so called from the flowers appearing when the Shad-flies first rise from the water in the month of May.

The elegant white flowers of this pretty tree (for it rises to the height of twenty feet) adorn the banks of our rivers and lakes and enliven the surrounding woods, breaking the monotony of their verdure by the contrast of its snow-white pendent buds and blossoms. The branches of the Shad-bush are somewhat straggling ; the leaves of a bluish-green, ovate and serrated, white underneath ; the fruit of a dark red, sweet and pleasant. This tree loves gravelly banks, and may usually be found near rivers. It is the tallest of the June-berries ; it thrives well under garden culture and is a pretty object when in flower, but not so much so as the next variety, *Amelanchier Canadensis*, var. *oblongifolia* which is a tall, upright, slenderly-branched pyramidal bush, rarely exeeding twelve or fifteen feet in height ; it is very symmetrical in its growth, forming a fine compact pyramid, covered early in the month of May with an abundance of drooping racemes of elegant white flowers, sometimes tinged with pink ; the blossoms come somewhat before the tender silken leaf-buds unfold. The foliage is delicately and sharply cut at the margins of the thin, ovate, oblong leaves, which are soft, silky and folded together ; at first they are of a reddish-bronze, but they take a bright tint of green when more mature.    The flowers are on slender foot-stalks, petals narrow and wavy.    The calyx remains persistent, as in the Pear and Apple.    The fruit of this pretty June-berry is small ; when ripe it is of a pink or rose colour ; sweet and juicy but somewhat insipid ; not so nice as another form which is known in some places by the name of Sheep-berry.    This forms a handsome bush about ten feet high, the flower and fruit larger than the former; the berries dark red, almost purple when ripe in July, with a pleasant nutty flavour.    Open thickets on the sides of ravines on the Rice Lake plains were favourite localities for the Sheep-berry.    Another dwarf June-berry, not more than five or six feet high or less, grows in the sandy flats on these same plains.    This is a pretty low shrub with green-ish-white racemes of flowers and oval leaves, fruit dark purplish-red, sweet but the berries are small, not larger than currants ; the bark of the branchlets of this little June-berry, is dark red, and the leaves are very downy underneath, the fruit is ripe in July and August about the same time as the Huckleberries.

### DWARF CHERRY—SAND CHERRY.—*Prunus pumila* (L).

The Dwarf Cherry, more commonly known as Sand Cherry, is chiefly found on light, sandy lands ; it is a low, bushy shrub, from eighteen inches to two feet in height : the slender branches are inclined to trail upon the ground, sometimes rooting ; the centre stem is more upright.    This little cherry has a pretty appearance when covered with the clusters of small, white, almond-scented blossoms, which on short slender foot stalks spring, in twos or fours, from the base of the small pale-green leaves that clothe the reddish-barked branches ; the fruit, not exceeding the size of a common pea, is purplish-red, without bloom on the surface.    The Sand Cherry abounds on light plain-lands ; it is the smallest of the wild Cherries, and is far more palatable than the fruit of some of the larger trees of the Genus.    In flavour it partakes more of the nature of the Damson or Plum.    Possibly under cultivation the fruit might be greatly improved in size and quality : and the plant is so pretty an object, whether in flower or fruit, that it would repay the trouble of cultivation in the garden as an ornamental dwarf shrub.    So eagerly is the fruit sought for by the Pigeons and Partridges, that it is difficult to obtain any quantity even in its most favoured localities.

### CHOKE-CHERRY.—*Prunus Virginiana* (L).

Very tempting to the eye is the dark-crimson, semi-transparent fruit, when fully ripe, of the Choke-Cherry, and not unpalatable, but so very astringent that it causes a painful contraction of the throat if many berries are eaten at one time, though some persons are not much affected by them, and will take them freely without any ill consequences. The bush is from eight to ten feet high, flowering abundantly and forming a pretty object from the profusion of long, graceful, pendulous racemes of greenish-white flowers which have an almond-like scent when fully blown. The leaves also have a pleasant, aromatic, bitter flavour like those of the Peach and Almond, and form a good flavouring, resembling Ratafia ; when boiled in milk for puddings and custards one or two are sufficient, and may be removed when the milk has boiled. This flavouring is harmless and pleasant, and easily obtained.

The Choke-Cherry never reaches to the dignity of a tree like the Wild Black and Wild Red Cherry of the woods, but forms a pretty flowery shrub of straggling growth. It blossoms in June and ripens the fruit in August. In both stages, of flower and fruit, it is very ornamental, and may be introduced with advantage to the shrubbery— but so tempting are the ripe berries to the smaller fruit-loving birds that it is soon stripped of its rich crimson load of pendent fruit. The Cedar or Cherry-Birds are sure to find out the bush and visit it in flocks till they strip it entirely, leaving the ground below strewed with the berries that have been shaken off : possibly the Ground Squirrels and Field-mice thus come in for a share of the spoils.

PRICKLY ASH.— *Xanthoxylum Americanum*, (Mill.)

This is a handsome shrub with glossy pinnate leaves, the valuable qualities of which are hardly sufficiently known and appreciated by those who know it only for its ornamental appearance, when the crimson cases that envelop the black shining seeds appear in clusters between the bright green leaves. The leaflets are in five pairs, with one terminal, from an inch to two inches in length, serrated at the edges, pointed, of a lively bright green, very glossy on the surface. The stem and branches straight, covered with whitish grey bark ; the branches set with stout woody prickles, which also extend along the mid-rib on the underside of the leaves. The flowers are yellowish green, in close set clusters, appearing before the leaves. The fruit is a round, hard, shining bead-like berry, on a little thready stalk, two in each pod, at first a bronzed green, deepening to deep crimson when ripe, opening and shewing the dark glossy seeds. The whole plant is highly aromatic, especially the cases that enclose the seeds, which, when rubbed between the fingers, emit a strong pungent odour, like the scent of Orange-peel.

The root, bark, leaves, and fruit, are bitter, pungent and aromatic. The root and bark are used in dyeing yellow : they are also used medicinally in extract for Agues and Intermittent Fevers.

Though its most usual locality is on the banks of streams and in low wettish ground, it will also thrive and increase rapidly on dry soil, and on account of its stout woody stem it seems well suited for hedges. The Prickly Ash will grow both from seed and by shoots sent up from the roots. The fruit is ripe in August and September. The dry seed-pods are in great request by smokers, who mix them with tobacco and regard the fine spicy scent as a great luxury when they can obtain the berries from the Indians.

The following valuable remarks on the medicinal uses of this interesting shrub were copied for me by my late much-valued friend, Dr. Low, of Bowmanville, from the Journal of Materia Medica, No. XII., December 1859, by Dr. Charles Lee, on the Medicinal Plants of North America :—

" The ' Prickly Ash ' is known also by the name of 'Yellow-wood.' The bark contains a fixed volatile oil, resinous colouring matter; gum and a crystalizable substance. The berries contain a large amount of oil, one pound yielding four fluid ounces when treated with alcoholic ether. The Prickly Ash is employed as a remedy for affections of the spine, marrow, and vascular system. The active properties consist of an ethereal oil, like oil of turpentine, it is decidedly stimulant in languid cases of the nervous system.

" In Asiatic cholera, during the years 1848-50, it was used with great success by American physicians in Cincinnati : it acted like electricity, so sudden and diffusive was the effect on the system.

" In the Summer complaint of young children it is also used with great success. The following is an excellent receipt for that disease among children :—

" Rhubarb root, Colombo, Cinnamon—of each 1 drachm ; Prickly Ash Berries, 3 drachms ; Good Brandy, half a pint. Add the bruised articles to the brandy, shaking them for three or four days occasionally. The dose for a child of two years old is a teaspoonful thrice a day in sweetened water. Where any swelling of the body is apparent, equal parts of the tincture of Prickly Ash Berries and Olive Oil is of great use rubbed in over the abdomen. In typhus and typhoid fevers, the value of this tincture is very great. A teaspoonful diluted with water may be given, in cases of great depression and prostration, every twenty minutes ; it is also used most successfully in chronic rheumatism."

I make no apology for introducing the above, thinking it may prove a valuable receipt.

Another of our lovely creeping forest evergreens is the

CREEPING SNOW-BERRY.—*Chiogenes hispidula*, (T. & G.)

This interesting little plant forms beds in the spongy soil of the damp cedar swamps, spreading its matted trailing branchlets over the mossy trunks of fallen trees. The foliage is dark green—very small—and myrtle-like in texture, hard and glossy. The flowers, which are solilary in the axils of the leaves, are not very showy; they are bell-shaped and four-cleft at the margin, greenish-white in colour. The berry is pure white and waxy, and lying on the deep green mat of tiny evergreen leaves, has a charming effect.

*Chiogenes hispidula* belongs to the Heath family, and grows in cool peat bogs and mossy mountain woods, in the shade of evergreens; the whole plant has the aromatic flavour of the Teaberry or Aromatic Winter-green, *Gaultheria procumbens.*

HUCKLEBERRIES—BLUEBERRIES.

Several varieties of this useful and agreeable fruit are spread all over the country, even to the farthest Northern and Eastern portions of the now widely extended Dominion. Many of the species are hardy, and will bear the severity of almost Polar cold, and will flourish in the poorest soil. The commonest to be met with are the large Blueberries, *Vaccinium Pennsylvanicum*, *V. Canadense* and *V. corymbosum*, which abound in the Oak-openings, in swamps, and on the stony islands of our back lakes.

DWARF BLUEBERRY—*Vaccinium Pennsylvanicum*, (Lam.)

Is the earliest to ripen its large sweet berries. The flowers, which are delicate waxy bells, appear early in May, and are with the young leaves pinkish in colour. The leaves are lanceolate with serrated margins, smooth and shining on both sides. The berry is ripe early in July, and is the earliest Blueberry brought to the market.

This is a low bush, one to two feet high, found growing in woods and on the borders of swamps.

CANADA BLUEBERRY.—*Vaccinium Canadense*, (Kalm)

Is a low shrub with downy branches and leaves, very similar to the above, but generally smaller, and with shorter greenish flowers, striped with red; the leaves are not serrated at the margin, and the fruit is not quite so early. It generally grows in damper situations.

SWAMP BLUEBERRY.—*Vaccinium corymbosum*, (L.)

This is a large handsome shrub, five to eight feet high, found in many varieties growing in swamps. The corolla is larger than either of the above and of a purer white.  The leaves ovate and entire, and slightly pubescent.  The rich berries begin to ripen in August, and are the latest of the season.

These pretty shrubs, loaded with their luscious berries, may be found on all dry open places.  The poor Indian squaw fills her bark baskets with the fruit and brings them to the villages to trade for flour, tea, and calico, while social parties of the settlers used to go forth annually to gather the fruit for preserving, or for the pleasure of spending a long Summer's day among the romantic hills and valleys ; roaming in unrestrained freedom among the wild flowers that are scattered in rich profusion over those open tracts of land, where these useful berries grow.

These rural parties would sometimes muster to the extent of fifty or even an hundred individuals, furnished with provisions and all the appliances for an extended pic-nic.

Many years ago, when the beautiful Rice Lake Plains lay an uncultivated wilderness of wild fruits and flowers shaded by noble, wide spreading Oaks, silver Birches and feathery Pines, an event occurred that excited great interest in the neighbourhood, and for miles around, the excitement even penetrating to distant settlements on the Otonabee, then the border-land of civilization, North of the Great Lakes.

It was in the month of July, 1837, that a large party of friends and neighbours near Port Hope agreed to make a pic-nic party, to gather Huckleberries and pass a pleasant Summer day on the Rice Lake Plains.

They made a large gathering in waggons and buggies and on horseback.  Among the children belonging to the party was a little girl about seven years of age, a bright, engaging child,  By some accident this little one got separated from her family among the bushes, and they, supposing that she had gone forward with some of their near neighbours and friends, started for home, feeling no uneasiness until it was discovered that little Jane was not among the returned party, and that no trace of her could be found.

Then came the stunning conviction that the child was lost—left alone to wander over that  pathless wilderness in darkness and solitude, perhaps to fall an unresisting prey to the Bear or the Wolf, both of which animals at that distant period roamed the hills and ravines of those plains in numbers, unchecked by the rifle of the sportsman or the gun of the Indian hunter.

A few cleared spots there were : but these were miles apart, and it was not likely that the timid child would find her way to any of the distant shanties, so that no reasonable hope of the child finding shelter for the night could be entertained. Under so sad a loss, the distress of the bereaved parents may easily be imagined. Their agonizing suspense, their hopes and their fears, found a ready response in every kind and feeling heart.

No sooner was it known that a young child was lost, than hundreds of persons interested themselves in the discovery and restoration of little Jane Ayre. The people came from their farms ; they poured out from towns and villages, from the borders of the forest ; wherever the tale was told came men in waggons, on horseback, and on foot, to scour the plains in every direction.

The Indians, under their Chief, Pondash, came under promise of a liberal reward if they found the child. Day after day passed without tidings of the lost one. As night came on each party returned, only to say the child was not found, and hope began to fade away in all hearts. It still lingered however in that of the father.

It was now Thursday, and it was on the evening of the previous Saturday that the little girl had been lost. The chances were indeed remote that she would be found, or if found, that she would be a living, breathing child.

However, about noon on the Thursday a horseman was seen riding at full speed towards the farm, followed by a crowd that thronged the road. The lost child was found ! Alive or dead ? There was a stop, a pause, in the pulsation of the woe-worn heart of the mother. Could it be that after five days of famine and wandering, exposed to the rain and dews, and the sun's hot rays, that she should behold her child alive once more ? Yes, it was even so, and He who tempers the rough wind to the shorn lamb and shelters the unfledged nestling of the wild birds, had been her guard by night from the wild beasts and her shield by day from the elements. No harm had befallen the young wanderer, save what naturally arose from exhaustion and fear in her unusual position.

Each night she had lain down, and, sheltered by a fallen Pine tree, had slept as soundly as if on her own little bed at home. The first night a drenching thunder-shower had wetted her clothes, and she had lost her shoes in the grass and she had not cared to seek for them ; her face was much sunburnt, and she said each day she had heard voices in the distance, but her fear of strangers, and especially of Indians, had made her conceal herself. One thing was remarkable— hope and trust in her father had never deserted her young heart. She said, she knew that he would never cease to look for her till she was found. It was with the hope of seeing that dear face that she came from her hiding place and stood upon the log and looked about her, and was fortunately discovered by one of the searchers whom she knew by sight—and then what a cry of joy arose, such as those wild plains had never echoed before, " The Child ! The Child ! "—it reached the father's ears, though distant far from the spot, and he scarcely believed yet, for joy, till she was placed warm and breathing in his arms. The crowd instinctively drew back for a space and left the father and child clasped in each other's arms. Many a manly cheek was wet that day when they saw the childish face, thin and wan as it was, nestling in the father's arms, her thin browned hands clasped about his neck as if no power on earth should part them again.

Surely the father might have cried out in the fulness of his heart " Rejoice with me, my friends, for this my lamb was lost and is found !"

Years have passed away, and little Jane has long been a wife and happy mother, and no doubt has often told her children the tale of her being lost on the Rice Lake Plains, and pointed them to the gracious Father in Heaven, who kept her under the shadow of His wing during those days of danger, fear, and famine.

The plains are now cultivated in every direction; the Huckleberries are fast disappearing and will have to be sought for elsewhere.

### Frost Grape—*Vitis cordifolia*, (Mx.)

Those deep, embowering masses of foliage : those verdant draperies that fall in such graceful, leafy curtains from branch to branch, roofing the dark shady recesses of our wooded lakes and river banks : those light feathery-clustered blossoms that hover like a misty cloud above the leafy mass, giving out a tender perfume as the breeze passes over them—like sweet Mignonette—those are our native vines, our Wild Grapes.

Yon tall dead tree, that stands above the river's brink, is wreathed with a dense mantle of foliage not its own. The changing hues of the leaves. the deep purplish clusters of fruit, now partially seen, now hidden from the view, have given a life and beauty to that dead unsightly tree.

The ambitious parasite has climbed unchecked to the very top-most branch, and now flings down its luxuriant arms, vainly endeavouring to clasp some distant bough ; but no, the distance is beyond its reach, and it must once more bend earthward or in lieu of better support, entwine its flexile tendrils in a tangled network of twisted sprays, leaf-stalks, and embowering leaves and fruit.

The fruit of the Frost-grape—our Northern grape-vine—is small. The berries, round blue or black with little or no bloom, very acid, but edible when touched by the frost, and can be manufactured into a fine jelly and good wine of a deep colour and high flavour. Whole islands in the Trent and Rice Lake are covered with a growth of this native Grape. There is not a lake in Canada but has its "Grape Island," and many persons cultivate the plants about their dwellings over light trellis work, under which circumstances they will yield an abundance of fruit. It is also very useful to conceal unsightly objects, as out-houses An old pine stump can be converted into an ornamental object, by nailing cedar poles—fastened at the top—round it, and planting grape-vines around it, having first prepared a bed of good earth and large stones, to bank the lower part ; a few plants of the Wild Clematis intermixed with the Grape-vine and a sprinkling of Morning Glories, make a lovely pyramid and convert a defect into a charming object, during many months of the year.

The Wild Grape seems to flourish best, in its natural state, near the water, but will grow and flourish well in gardens where it is given the support of a trellis or in any suitable position where it can climb. I have even seen a dead tree specially planted for such a purpose.

### Fox Grape—*Vites Labrusca*, (L.)

This is the original of the cultivated Isabella Grape, which has long been introduced into our gardens and vineries as worthy of the attention of fruit-growers.

The leaves of this species are very densely woolly, covered, especially when young, with tawny, silky hairs ; the fruit is of a dark purple, of a musky flavour, whence its common name, Fox Grape.

This Wild Grape is found on the shores of Lake Erie, and to the Westward. From the improvement made by cultivation, in the size and quality of the Wild Fox Grape, we may perceive how much might possibly be done with others of our wild fruits, which, when introduced into our gardens would have the advantage of hardiness in bearing the severity of our climate, beyond that of exotics.

It seems reasonable to suppose that plants that are indigenous to a country, could, by due care, be brought to a state of higher perfection than when under a foreign sun and soil, and that the culture of wild plants would amply repay the cultivator. Attempts of this kind are rarely made or persevered in, so that the result is not often satisfactory : either the process is thought to be too slow, or we despise as common, that which is within our reach, valuing that which is more costly above what is easily obtained ; whilst we eagerly spend our money to obtain a foreign species, which may possibly have been originally taken from

our native woods and wilds to a foreign country, there cherished and
cared for, improved by cultivation, and returned to us increased in
value.   It would greatly enhance the pleasure of cultivation if we were
ourselves, able to show native flowers and shrubs and fruits, rendered
equal to the imported kinds by our own culture.

We might compare these wild plants to the neglected children of
our poorest classes.   In the degradation arising from their uncared for
state they become as moral weeds in the great garden of life, neglected
and passed by, left to run wild, and shunned ; but remove these children
to a more genial atmosphere ; let them be taught the value of their souls,
for which so great a price was paid by their Redeemer ; let them be
clothed and fed, and cared for, made to feel that they are not despised
in the eyes of their fellow men ; then their useful qualities brought into
action, and their   vices and evil passions controlled, like the wild
plants, they will rise in value, and beauty, and usefulness, becoming
precious trees bearing fruit to the glory of Almighty God—sought out
and desired of all men.   Who will cultivate and improve this garden of
human growth ?   Must it continue a wilderness, rank and injurious, full
of deadly poisons and unripe, crude and bitter fruits ; while within it,
choked and hidden from view, are the germs of usefulness, beauty, and
happiness, that only require the better soil, the fostering care, and
gladdening sunshine of christian love and kindness, to make them what
their Creator would have them all to be ?   Truly "the harvest is great
but the husbandmen are few."

Allusions to the grape-vine and vineyards are of frequent occurrence
in Scripture.   Many and beautiful are the passages where the ancient
church is symbolized by the poetical figures of the vine and the vineyard.   How touching is the appeal made by the prophet to the rebel-
lious and idolatrous people in the fifth chapter of the book of Isaiah.

"And now, O inhabitants of Jerusalem, and ye, men of Judah,
judge I pray you betwixt me and my vineyard.   What could have been
done more to my vineyard, that I have not done in it ?   Wherefore when
I looked that it should bring forth grapes, brought it forth wild grapes."

Beautiful are the allusions made in the song of Solomon, in his
invitation to the beloved to go forth to the garden he had planted.
"The fig tree putteth forth her green figs, and the vine with the tender
grapes give a good smell.   Arise my love, my fair one, and come away."

"Let us get up early to the vineyards ; let us see if the vines
flourish, whether the tender grapes appear."

Probably the culture of the vine was among the earliest labours of
the husbandman, and must have been of most ancient usage, the first

work enjoined by the Almighty Creator when he placed man in the Garden of Eden—which was most likely a large and fertile tract of country already enriched with every tree, and herb, and flower, that would prove useful for the support of life and enjoyment. Adam was instructed by his Maker to till the ground and dress it and keep it.

This employment was ordained for health and pleasure, not for toil or weariness. This last condition arose when sin had marred the fair beauty of God's world and the sin-smitten earth no longer yielded its spontaneous fertility as in the day when sinless man first stood in his innocence on the then unpolluted earth, a fearless being in the presence of a holy God.

The vine which might have formed a delightful portion of man's food in the Edenic garden, must from henceforh yield its luscious grapes only by care and labour. The wild vines must be pruned and trained and kept free from noxious weeds and hurtful insects ; they were no longer the fruit of the Lord's vineyard. Who can tell but that our wild Canadian Frost and Fox Grapes may not be the degenerated seed of the wild vines of that land of the east, into which Adam and Eve were banished.

Travellers in Palestine still speak of the luxuriant Grape-vines flinging their clusters of fruit and sweet-scented blossoms over the terraced steeps of rocky ravines, filling the air with perfume ; but the vines are all wild now and uncultivated. They want the careful hand of the vine-dresser and husbandman to train them. Type of the wasted inheritance of the ancient people, and of a degenerate priesthood.

Has the Christian church no careless vine-dressers ; are there no vines bringing forth wild grapes ; no briars and thorns that come up to choke the Lord's vineyard, till it becomes an unfruitful wilderness ?

BLACK HAWTHORN—PEAR THORN—*Cratægus tomentosa*, (L.)

Canada has many species of Hawthorn ; but not the fragrant flowering May of the English hedgerows, associated in the minds of Old Country people, with the pleasant Spring days and bowery lanes of their childhood, when, as old Herrick tells us " Maids went maying." But even now in Merrie England, the May-queen's reign is over, in spite of poets' songs.

LAMENT FOR THE MAY-QUEEN.

No Maiden now with glowing brow
Shall rise with early dawn,
And bind her hair with chaplets rare
Torn from the blossomed thorn.

No lark shall spring on dewy wing
  Thy matin hymn to pour,
No cuckoo's voice shall shout "rejoice !"
  For thou art Queen no more.

Beneath thy flower-encircled wand,
  No peasant trains advance ;
No more they lead with sportive tread,
  The merry, merry dance.

The Violet blooms with modest grace
  Beneath its crest of leaves,
The Primrose shows her paly face ;
  Her wreaths the Woodbine weaves.

The Cowslip bends her golden head,
  And Daisies deck the lea ;
But ah, no more in grove or bower,
  The Queen of May we'll see.

Weep, weep then virgin Queen of May,
  Thy ancient reign is o'er ;
Thy votaries now are lowly laid,
  And thou art Queen no more.

The Pear Thorn is one of the finest of our native species, it often rises to the height of from fifteen to twenty feet with a stout rough-barked stem. When in flower it forms a fine ornament to our open woods and thickets, for it is not found in the depths of the forest ; but at the open edges of woods, more especially it will be found along the banks of rivers and creeks. The flowers are much larger though less delicate in scent than the English Hawthorn. The leaves are thick and tough, but smooth and shining, unequally toothed, ovate-oblong ; thorns, long sharp and slender. The white cup-shaped flowers with dark anthers grow in handsome corymbs, many-flowered on the summits of the sprays. The fruit is large, round and of a bright scarlet or orange.

#### SCARLET-FRUITED THORN—*Cratægus coccinea*, (L.)

Is no less ornamental than the former, it also forms a fine high flowering bush ; the fruit is of a pleasant acid and of a fine bright scarlet, the leaves are thin, partly lobed and sharply cut at the rounded margin. This thorn grows tall and slender in close thickets and shade, but seems to prefer open ground and plenty of sunshine, when it forms a lovely small, compact, tree and flowers abundantly ; the fruit is not so large as in the last species, and is of a deeper red colour.

The English White Thorn, *Cratægus oxyacantha*, (L.) in some situations grows beautifully, but is apt to dwindle and become mossy and gnarled in unsuitable places where it is neglected.

PLATE V.

I. FALSE FOXGLOVE (*Gerardia quercifolia*).

II. LESSER FOXGLOVE (*Gerardia pedicularia*).

A most perfect specimen of the English White-thorn may be seen at Port Hope on the lawn at the residence of C. Kirkhoffer, Esq., at the western side of the town ; it was in full flower when I saw it, and formed one of the most beautiful objects I ever saw, it was worth going miles to look upon it, and to inhale the sweetness of its abundant white blossoms.

There appears to have been little attempt made to cultivate our Hawthorns as hedge plants, though one might naturally suppose that such would have been adopted in places where the difficulty and expense of obtaining rail-timber is now being sensibly felt by the farmer. The Cedar and Hemlock are largely used for garden enclosures. Why not try the Hawthorn also ?

### Small Cranberry—*Vaccinium Oxycoccus* (L.)

> There's not a flower but shews some touch
> In freckle, freck or stain,
> Of His unrivalled pencil.—*Hemans*

There is scarcely to be found a lovelier little plant than the common Marsh Cranberry. It is of a trailing habit, creeping along the ground, rooting at every joint, and sending up little leafy upright stems, from which spring long slender thready pedicels, each terminated by a delicate peach-blossom-tinted flower, nodding on the stalk, so as to throw the narrow pointed petals upward. The leaves are small, of a dark myrtle-green, revolute at the edges, whitish beneath, unequally distributed along the stem. The deep crimson smooth oval berries are collected by the squaws and sold at a high price in the Fall of the year.

There are extensive tracts of low, sandy, swampy flats, in various portions of Canada, covered with a luxuriant growth of low Cranberries. These spots are known as Cranberry Marshes, and are generally overflowed during the Spring ; many interesting and rare plants are found in these marshes, with Mosses and Lichens not to be found elsewhere, low evergreens of the Heath family, and some rare plants belonging to the Orchidaceæ such as the beautiful Grass Pink (*Calogogon pulchellus*) and *Calypso borealis.*

Not only is the fruit of the low Cranberry in great esteem for tarts and preserves, but it is considered to possess valuable medicinal properties, having been long used in cancerous affections as an outward application. The berries in their uncooked state are acid and powerfully astringent.

This fruit is successfully cultivated for the market in many parts of the Northern States of America, and is said to repay the cost of culture in a very profitable manner.

J

So much in request as Cranberries are for household use, it seems strange that no enterprising person has yet undertaken to supply the markets of Canada. In suitable soil the crop could hardly prove a failure, with care and attention to the selection of the plants at a proper season.

The Cranberry belongs to one of the sub-orders of the Heath family *(Ericaceæ)*, nor are its delicate pink-tinted flowers less beautiful than many of the exotic plants of that Order, which we rear with care and pains in the green-house and conservatory; yet, growing in our midst as it were, few persons that luxuriate in the rich preserve that is made from the ripe fruit, have ever seen the elegant trailing-plant, with its graceful blossoms and myrtle-like foliage.

The botanical name is of Greek origin, from *oxus*, sour, and *coccus*, a berry. The plant thrives best in wet sandy soil and low mossy marshes.

WILLOW-LEAVED MEADOW-SWEET.—*Spiræa salicifolia*, (L.)

Frederic Pursh, in his North American Flora—a valuable work but little referred to—gives no less than seven different species of this Genus Spiræa as natives of Canada ; the description of two or three will be sufficient for the present limited work on the indigenous shrubs of this portion of the Dominion. Of the white flowered species, *Spiræa salicifolia*, the Willow-leaved Meadow-sweet is the most commonly met with, and is often found in gardens and shrubberies. It is a pretty, graceful shrub, with clustered feathery panicles of white or pale waxy-pink flowers, which are terminal on slender branches ; the leaves long, narrow and thin, of a pale green, serrated on the margins. Our Spiræas will not only bear removal to the garden, but flourish luxuriantly under cultivation. The only objection to their introduction to our borders is that they are apt to become too intrusive, by throwing up many suckers, which have to be rooted out.

A very slender variety, with simple wand-like stems and terminal spikes of small white flowers, may be found growing among the cracks and fissures of the rocky shores of Stoney Lake and its numerous islets, rooting in sterile spots among the few wild grasses that find nurture in the scanty mould that is lodged in such crevices. This delicate little shrub may be found in flower all through the hot months of July and August. The Spiræas belong to the Rose family. The popular name, Meadow-sweet, seems hardly appropriate to our pretty shrub, as it has very little fragrance. But this name for the whole Genus is taken from the beautiful and odoriferous British species, *Spiræa Ulmaria.*

Hard-hack—Rose Coloured Spiræa.—*Spiræa tomentosa*, (L.)

Of the several pretty shrubs belonging to the Genus *Spiræa*, which have been introduced into cultivation, none deserve a place in our gardens more decidedly than the above. It is a beautiful shrub, growing in wild profusion in swamps and on the rocky shores of our small inland lakes. It is about four feet high, with slender, wand-like stems that rise from a woody root-stock, clothed with dark green, serrated, leathery leaves, which are smooth above, but very downy underneath. The flowers are of a fine rose-pink, in closely-flowered panicles, a little branching in the larger heads. The bark of the stem is red, and covered with whitish down.

While this elegant shrub is chiefly found near water, it seems to prefer a gravelly or rocky soil for its habitation.

Purple Flowering Raspberry.—*Rubus odoratus*, (L.)

In English gardens our beautiful sweet-scented Raspberry is deemed worthy of a place in the shrubberies, but in its native country it is passed by and regarded as of little worth. Yet what can be more lovely than its rose-shaped blossoms, from the deep purplish-crimson bud wrapped in its odorous mossy calyx, to the unfolded flower of various shades of deep rose and paler reddish lilac. The flowers derive their pleasant aromatic odour from the closely-set coating of short bristly glandular hairs, each one of which is tipped with a gland of reddish hue, containing a sweet-scented gum, as in the mossy envelope of the Moss-Rose of the garden. These appendages, seen by the aid of a powerful microscope, are objects of exquisite beauty, more admirable than rubies and diamonds, living gems that fill us with wonder while we gaze into their marvellous parts and glorious colours.

All through the hot months of June, July and August, a succession of flowers is put forth at the ends of the branches and branchlets of our Sweet Raspberry—

"An odorous chaplet of sweet summer buds."

The shrub is from two to five feet in height, branching from the woody perennial root-stock ; the leaves are from three to five lobed, the lobes pointed and roughly toothed. The leaves are of a dullish green, varying in size from several inches in diameter to mere bracts. The blossoms are often as large as those of the Sweet-Briar and Dog-Rose, but when first unfolded are more compact and cup-like. The fruit, which is popularly known by the name of Wild Mulberry, consists of many small red grains, somewhat dry and acid, scarcely tempting to the palate, but not injurious in any degree. The shrub is more attractive for its flowers than its insipid fruit. We have indeed few that are more ornamental among our native plants than this *Rubus.* Canada possesses many attractive shrubs that are but little known, which flourish year after year on the lonely shores of our inland lakes and marshy Beaver-meadows, unnoticed and uncared for in their solitary native haunts.

Closely resembling the Purple Flowering-Raspberry, is the White Flowering-Raspberry, *R. Nutkanus,* (Mocino), the chief difference being in the colour of the flowers and the shape of the petals, which, in the latter species, are of a lovely pure white and oval in shape. The whole plant is slightly smaller and less bristly. The fruit is very similar in both species.

WILD RED RASPBERRY.—*Rubus strigosus,* (Michx.)

The wild Raspberry springs up spontaneously all over Canada. In the forest, in newly made clearings after the fire has passed over the ground, on every upturned root, in the angles of the snake-fences, and on every waste and neglected spot, the Raspberry appears and takes possession of the land. Truly this useful and palatable fruit proves a blessing and a comfort in various ways to the poor, as well as a wholesome, welcome luxury to the richer inhabitants of our towns and villages. During the fruiting season the women and children are enabled to supply many household wants by the sale of the red and black Raspberries ; even the little ones are made to contribute their small mite of labour, and may be seen in large parties going out with tins and sundry small vessels to the Raspberry grounds. Wild rugged spots that have been abandoned by the farmers ; worthless for the growth of roots and grain. He cannot look beyond and see that with Our bountiful Provider there are no waste places. He who fed the wandering multitude with Manna in the thirsty desert, and brought forth springs of water from the flinty rock, can give fruits to satisfy the wants of his children in the Canadian wilderness. The wild berries are shared by God's humbler gleaners the small animals, and flocks of birds ; and even the insects all come to this table that is spread abroad for them and us ; " and something gathers up all fragments and nothing is lost."

The fruit of the common Red Raspberry begins to ripen early in the month of July, just about the time that the Strawberry ceases to be plentiful. The flowers are not very ornamental, whitish, but not clear white, rosaceous in form. The berry ripens very soon after the fading of the flowers.

The colour of the fruit of the common Raspberry is of a light red, changing with maturity to a dark crimson. The bush is upright—not very prickly. The leaves have from three to five leaflets, greyish or dull green, wrinkled and veiny, whitish underneath ; leaflets serrate, unequally lobed, pointed ; the fruit is juicy and acid, not as sweet as that of the

## BLACK RASPBERRY—*Rubus occidentalis*, (L.)

This species is distinguished from the above by its long arching flexile branches covered with purplish red bark, strongly hooked prickles and blackish fruit, very rich, firm and sweet. It loves to grow on hilly banks and upturned roots in the shade of the forest where it can send down its long flexible branches, which bear an abundance of berries long after the Red Raspberry has failed to yield a supply. Gray calls this Black Raspberry by the familiar name of Thimble-berry ; but it is the fruit of the Blackberry—*Rubus villosus*, (Ait.) that is commonly known by this name. The berries of the Blackberry are not hollow, nor do they, like the last, separate from the receptacle ; they are conical, sweet and luscious to the taste, in quality astringent, but not unpleasantly flavoured. The berries ripen in August ; the foliage is veiny, coarse, with strong red prickles, the stems strongly armed and covered with a dark-red bark, which with the root is highly astringent and used both in the form of a tea and syrup in cases of Dysentery and Summer-complaint. The fruit in syrup is also considered medicinal and useful in similar complaints. A very pretty, ornamental, low, creeping, shrubby plant is the

## SWAMP BLACKBERRY—*Rubus hispidus*, (L.)

The branches, very strongly armed with hooked prickles, are long and slender, extending two or three feet over the ground, leaves of three leaflets, bright varnished green, rounded at the ends, more in form like those of the Strawberry ; flowers rather large, very delicately tinted with pinkish or else white, like a small, single, Briar Rose. This low Blackberry seems to love rocky ground, creeping among stones and rooting in the black mould in the crevices ; the fruit is blackish-purple and pleasant to the taste.

## THE SWAMP-BERRY—*Rubus triflorus*, (Richardson.)

Is a pretty low trailing plant bearing somewhat insignificant white flowers, and ruby-coloured juicy acid fruit ; it ripens about the same time as the wild Strawberry, and the plants are seen running among the wild grasses and Strawberry vines, conspicuous by the lighter green leaves, which grow in compounds from three to five, coarsely, doubly serrate, and sharply pointed ; the flowers in small bunches of three. Like that of all the Genus, the fruit is perfectly wholesome.

### Early Wild Rose—*Rosa blanda*, (Ait.)

> " Nor did I wonder at the Lilies white,
> Nor praise the deep vermillion of the rose."—
>
> " The Rose looks fair, but fairer we it deem,
> For that sweet odour which in it doth live."—*Shakespeare.*

The Early Wild Rose, *Rosa blanda*, is hardly so deeply tinted as our Dwarf Wild Rose, *Rosa lucida*, but both possess attractions of colour and fragrance ; qualities that have made the Rose the theme of many a poet's song.  In the flowery language of the East, Beauty and the Rose seem almost to be synonymous terms.  The Italian poets are full of allusions to this lovely flower, especially to the red Damask Rose.

A popular song in the days of Charles I. was that beginning with the lines—

> " Gather your Roses while you may,
> For time is still a flying,
> And that same flower that blooms to-day,
> To-morrow may be dying."

The leaves of *Rosa blanda* are pale underneath ; leaflets five to seven ; flowers blush-pink : stem not very prickly ; fruit red and round ; the bush from one to three feet in height.

### Dwarf Wild Rose—*R. lucida*, (Ehrh.)

Is widely diffused over Canada ; it is found on all open plain-lands, but shuns the deep shade of the forest.  The bark is of a bright red, and the young wood is armed with bristly prickles of a greyish colour. When growing in shade, the half-opened flowers and buds are of a deep pink or carmine, but where more exposed in sunny spots, the petals fade to a pale blush-colour.  This shrub becomes somewhat troublesome if encouraged in the garden, from the running roots, which send up many shoots.  In its wild state the Dwarf Rose seldom exceeds three feet in height ; it is the second and older wood that bears the flowers ; the flower-bearing branches become almost smooth or only remotely thorny. The leaflets vary in number from five to nine ; they are sharply serrated at the edges, and smooth on the surface ; the globular scarlet fruit is flattened at the eye and is of a pleasant sub-acid taste.

This beautiful red-barked Rose grows in great profusion on the plains above Rice Lake, clothing large tracts of hill and dale, and scenting the evening air at dew-fall with its delicate fragrance.

The Swamp Rose, *Rosa Carolina*, (L.) is not uncommon ; it is often seen growing at the margins of lakes and rivers, and at the edges of stony islands ; it will climb, with the aid of supporting trees, to the height of eight and ten feet. The numerous and showy flowers are of a somewhat purplish tinge of pink and are borne in corymbs ; the leaves are whitish underneath. This rose is armed with stout hooked prickles below, on the old woody stem, but is smoother above ; the flowers are more clustered than in the other species.

The Sweet Briar is often found growing in waste places, and in thickets near clearings—no doubt the seed has been carried thither by those unconscious husbandmen, the wild birds and the Squirrels that feed upon the heps as they ripen. The leaves retain for some time their sweet fragrance, that is so delicious.

There is a delicate, pale-flowered Sweet Briar Rose, *Rosa micrantha*, (Smith) having small foliage and numerous blossoms ; stems low and branching and covered with hooked prickles, which has been found growing on the high Oak-hills in the township of Rawdon ; and which, I am informed, is not uncommon in similar localities in Western Canada.

Wax-work—Climbing Bittersweet—*Celastrus scandens*, (L.)

This highly ornamental climber, with its clusters of conspicuous berries, is a great adornment to open woods during the late autumnal months, and indeed all through the winter, twining round the stems of slender saplings of White Birch, Cherry, Ash, and Elm, not unfrequently clinging so closely to its supporter as to form an intimate union with the bark, its own smooth, slender stem, in serpent-like coils, forming graceful volutes round the column of the unfortunate tree which suffers from the close embrace that stops the free circulation of the sap in its upward ascent to the branches. The Climbing Bittersweet is a rapid grower, and consequently a bold enemy that takes forcible possession of any young sapling which comes within its reach ; a very Old Man of the Sea that, once fixed, no blast of wind can shake off. But while we take the liberty of railing at the unconscious intruder, we must not omit to dwell upon its good qualities. Its brilliant scarlet arils (coverings of the seeds) and orange fruit that in profusion ornament the tree about which it twines, enliven the dull woods at a season when bright tints have ceased to charm the eye, and all the glories of Maple, Cherry, Birch, Ash and Beech lie mouldering on the ground at our feet, we may then look upwards to some slender silvery-barked Birch or grey Butternut and admire the gorgeous scarlet festoons that hang so gracefully among the naked, leafless branches. The plant, too, is very attractive in its spring verdure. The delicate leaves are ovate-oblong, narrowing towards the point, finely serrated, alternate ; the flowers in raceme-like clusters are yellowish-green, followed by round, smooth, berry-like pods which deepen, as the summer advances, from yellow to orange and from orange to bright scarlet. When the seeds are ripe the pod divides, and the segments curl back and disclose the three-celled three-valved berry, which has, in each cell, one or two hard yellow seeds, covered with a thin coating of scarlet pulp which is called the aril ; this is acrid and burning to the taste. The Indians make use of the acrid juices of this plant, from the inner bark of the root and the bruised berries, to compound an ointment which is stimulant and healing for old sores, chilblains, and disorders of a similar nature. In country places in England, I have seen the berries of the Black Bryony boiled down with lard, for an application to chilblains which had a similar effect to the Indian Bittersweet salve. The Indians also apply this remedy to burns. The inner bark also is used as an orange dye by the natives.*

There are several species belonging to this Order found in Canada ; but though very ornamental in cultivation as shrubs, none are climbing like our forest Bittersweet, or give such enduring winter ornaments to our houses. Mixed with the branches of Spruce, Hemlock, and Balsam Fir, it forms a substitute at Christmas in our churches for the bright, glossy leaves and red berries of the English Holly.

The Greek name of this ornamental shrub is derived from a word meaning,—latter season, on account of the fruit remaining persistent through the winter.

If the Bittersweet were planted in shrubberies, or among trees in plantations, it would become an enduring ornament and enliven the dulness of our Canadian landscape with its bright colours, during the long months of winter.

LABRADOR-TEA.—*Ledum latifolium*, (Ait.)

This is another of our medicinal shrubs, and was held in great repute among the lumbermen and the old backwoodsmen for its sanatory qualities, as a strengthener and purifier of the blood, and as being good for the system in various inward complaints. Some of the old settlers used a decoction of the leaves as a substitute for tea, approving of the resinous aromatic flavour. I was induced to try the beverage, but did not find it to my taste, though it was on the whole preferable to Hemlock tea, another favourite beverage among backwoodsmen. As a medicine no doubt it deserves the commendations bestowed upon it. Though I did not care for the decoction of the leaves, I was charmed with the beauty of the plant, when I first saw it growing on the banks of one of the lakes north of Peterborough. The whole aspect of this remarkable shrub is most interesting. In height it varies from two to

---

* The name Bittersweet is taken from the graceful English climber *Solanum dulcamara* (L.), from a fancied resemblance between the two plants. The English Bittersweet is sometimes found in Canada on the borders of swamps and in low woods, but is an introduced plant.

four feet, it is bushy in habit, but somewhat open and spreading ; the leaves are lanceolate, entire, very decidedly revolute at the margins, and clothed with a dense rust-coloured woolly felt beneath. The leaves are of a thick leathery texture, and dull brownish-green colour. The flowers are white, forming elegant umbel-like clusters at the summits of the slender sprays. As the heads of flowers are very abundant, this shrub forms a striking object, when seen growing in numbers, along the banks of lakes or in low flats, for it will flourish both on wet and dry situations, nor does it refuse to flower when brought into garden culture. It is a very ornamental object, deserving to be better known than at present seems to be the case. The leaves when bruised emit an agreeable resinous aromatic odour.

The roots of the Labrador-Tea are wiry and covered with a bitter astringent bark. Professor Lindley also mentions, in his Natural System of Botany, the astringent qualities of another member of the family *Ledum palustre*, (L.), a slightly smaller shrub with narrower leaves and oval instead of oblong pods ; the stamens too are uniformly ten instead of five and seven as in this species. *L. palustre* is found in the north of Europe and also in the far north in Canada.

WILD ROSEMARY.—*Andromeda polifolia*, (L.)

is another of our native shrubs which grows in peat bogs, and on the swampy margins of lakes, associated with Labrador Tea, the Pitcher Plant and the elegant Low bush Cranberry. The stems are from three to eighteen inches in height, and bear on the summits of the branches of the previous year the light purplish flowers, which are three to eight in number, on rather long pedicels and drooping in a one-sided raceme ; the stamens are ten in number and remain persistent on the dry berry-like capsule. The leaves are shining green above, glaucous-white beneath and have the margins so strongly revolute as to appear almost linear. This plant is said to have astringent and narcotic properties, and to give intoxicating qualities to liquids in which it is infused.

SILKY CORNEL—KINNIKINNIC.—*Cornus sericea*, (L.)

This species is the true Kinnikinnic of the Indians of Central Canada, the leaves and bark of which they use in the place of tobacco, or mixed with it. I have been told it is of an intoxicating quality. The bark is also used as a tonic and febrifuge. The berries are pale blue ; the flowers form flat cymes, and are greenish-white, the young bark is purplish. The bush grows to the height of eight to ten feet, in low damp rich ground forming dense thickets. There is a fine white silky fibre in the leaves, which may be seen by breaking the mid-rib across.　The thread is as fine and as frail as the delicate web with which some spiders envelop their eggs—too fine to be turned to any use.

The silken thread is not confined to this species alone, it exists in many other trees and plants.　In the nerves of several of the Dogwoods it is seen quite as conspicuously as in *C. sericea.*

PANICLED OR PRIVET-LEAVED CORNEL.—*Cornus paniculata,* (L'Her.)

This is a very pretty species of Dogwood, found abundantly on the Rice Lake Plains, on the high dry hills between the hamlets of Harwood and Gore's Landing.　The bush is not more than four or five feet high, with light branching sprays.　The pretty white flowers are borne in convex cymes or sometimes in panicles and are followed by snow-white berries.　The foliage is dark-green, often with a purplish-bronze tint ; the leaves are long and narrow ; the nerves, whitish, and the light veining distinctly marked ; the surface of the leaf is very smooth, but hardly shining.　This pretty shrub would be well worthy of being introduced into our shrubberies.

There are many other species of Dogwood which are common to our swamps and thickets, some reaching to the height of small trees, as the Flowering Dogwood, *C. florida,* which is held in great esteem in the United States, for certain medicinal qualities ; it has been used as a substitute for Peruvian bark in low fevers.　The Indians are said to extract a red dye from the roots.　The fruit of the Flowering Dogwood is scarlet : the flowers, with their showy creamy-white involucres, three inches across, are very handsome, and are produced abundantly in the month of June.　This very handsome shrub grows in Western Canada, where it sometimes becomes a tree and reaches to the height of twenty or thirty feet.　A great contrast is this stately species to the dwarf herbaceous creeping plant of our woods, *Cornus Canadensis.*

RED-OSIER DOGWOOD.—*Cornus stolonifera,* (Michx.)

There are few of the native species of Cornel that are more ornamental than the Red-Osier Dogwood ; the bright, crimson wand-like branches of which, even when stripped of their foliage, are an enduring ornament　Their rosy shadows mirrored on the surface of the smooth waters of lake or forest stream, enliven the landscape and delight the eye, when the beauty of the foliage of the surrounding trees and shrubs has been swept away before the autumnal frosts and wintry winds.

In Spring, and early Summer, the white, fragrant flowers, in crowded flat heads, adorn the low shores.　Later in the Fall, the blue berries on the bright red sprays are hardly less attractive. The fruit is unpalatable for man, but is eaten by some of the water-fowl that have their haunts in the lakes and inland waters. This species is the Kinnikinnic of the western and prairie Indians.

PARTRIDGE-BERRY—TRAILING WINTER-GREEN.—*Mitchella repens*, (L.)

Another of our pretty red-berried creeping forest-plants, is the Partridge-berry ; the flexile branchlets of this little plant spreading from the joints of the trailing stem, form a mat of dark green foliage, covering unsightly patches of decaying wood, roots, and stones with many a graceful wreath, as if Nature kindly placed them there to veil the rugged ground with grace and beauty, in the same way as the green Ivy clothes and adorns the mouldering ruin with its enduring verdure.

Each slender leafy spray of our pretty Winter-green is terminated by tubular, star-shaped, twin blossoms, which are divided at the margin into five sharply pointed segments ; white, sometimes slightly tinged with pink. The ovaries are united at the base of the flowers, and form one double-eyed round berry for each pair of flowers ; the interior of the flower-tube is hairy. The scent is sweet, faintly resembling that of the White Jessamine.

The berries remain persistent all through the Winter. They ripen to brilliant scarlet in the Autumn, and so continue till the return of Spring. Thus we may find fresh flowers, newly set fruit and the ripe berries all on the same plant. The small round leaves are veined with white, which gives a variegated look to their dark green surface.

The berries are mealy and insipid, but are eaten by the Indian-women and children as a dainty. These berries form food for the Wood-Grouse, our Canadian Partridge, and for the Woodchuck and other small quadrupeds that have their haunts in our forests and cedar-swamps. The elegant wreaths of dark variegated leaves and scarlet berries are sometimes used by Canadian girls as ornaments for their hair ; and I have seen white muslin evening dresses, trimmed with the sprays of this pretty evergreen, which had a charming effect, besides showing good taste and economy combined, in the fair wearers.

HIGH-BUSH CRANBERRY—AMERICAN GUELDER-ROSE—
*Viburnum Opulus*, (L.)

This fine shrub, with its large, loose cyme of white flowers, makes a goodly show during the month of June, mingling its snowy blossoms with the surrounding foliage of dark evergreens on the wooded banks of forest streams, and along the low shores of inland lakes and islands.

Not less attractive is it, when the full bunches of oval berries begin to ripen, first turning to amber, then brilliant orange-scarlet, and lastly, when touched by the frosts of Autumn, to a transparent crimson.    All through the winter you may see the bright ruby fruit upon the bushes, among the snow-clad branches, sometimes encased in crystal ice and magnified by the magic touch of hoar frost ; nor is the fruit of the High-bush Cranberry altogether useless to the Canadian housekeeper.    An excellent jelly is often made from the acid juice and pulp of the ripe fruit, when strained from the flat bony seeds, and boiled with sugar ; and though somewhat astringent, it forms an excellent sauce for roasted mutton or venison, and is useful as a fever drink mixed with water.

As a garden shrub this Viburnum is considered very ornamental, from its abundance of flowers and beautiful fruit.    It is no other than the fertile plant of the American Guelder Rose.    The cultivated Snow-ball Tree of our gardens is the same species, in which the fertile flowers have been suppressed and the showy sterile ones, which only appear in small numbers round the edge of the cyme in the wild plant, greatly increased in number by the skill of the horticulturist.    The *V. Opulus* is also indigenous to England ; and I remember finding the same flowering bush on the banks of a lonely pond in Reydon Wood, Suffolk, and recognized the High-bush Cranberry on the shores of the Otonabee River from its likeness to the shrub that had attracted my notice in my woodland rambles in England.

The foliage of the High-bush Cranberry takes a bronzed-purple hue, turning to a deep crimson in the Autumn.    The leaves are large, three-lobed and pointed.    The flowers are borne on wide-spreading peduncled cymes, having the central flowers very small but fertile ; the marginal ones are imperfect, being destitute of both stamens and pistils, but the corollas are disproportionately large and give the beauty to the flower clusters of this fine shrub.

The name Cranberry has been improperly applied to *Viburnum Opulus*, as it has no affinity with the low creeping Marsh Cranberry that most elegant and charming little plant, with its delicate graceful flowers, myrtle-like leaves, and pear-shaped ruby-coloured fruit.    Those persons who use the fruit as a preserve know little of the exquisite beauty of the plant itself.    To be admired, it should be seen in its native haunts growing among the soft peat-mosses of our marshes and bogs.    The wreaths of fine dark foliage, bearing the delicate pink waxy flowers on slender thready foot-stalks, and the large berries in every stage of progress—green, yellow, deep red and purplish red, resting upon the grey lichens and lovely cream-coloured peat mosses—produce an effect worth looking at.

The name of the Genus is supposed to be derived from the Latin word *vieo*, to tie, on account of the flexibility of the branches of some of the species. The word *Viburna*, in the plural, seems to have been applied by the ancients to all plants which were used for tying.

HOBBLE-BUSH—*Viburnum lantanoides*, (Michx).

This shrub would appear to be typical of the Genus, for the branches twine and twist most irregularly, and the lower ones are procumbent, often taking root where they touch the ground, whence the popular name. The flowers of this species somewhat resemble the last ; but are more cream-coloured, and appear earlier. The large hand-some leaves are round ovate, heart-shaped at the base, and, together with the young branchlets, are covered underneath on the veins and veinlets with tufts of brown down. The ovoid fruit is crimson, turning blackish, and although edible is not very pleasant.

MAPLE-LEAVED DOCKMACKIE—*Viburnum acerifolium*, (L).

is a low pretty shrub, not uncommon in open thickets and damp woods. The flowers are more delicate and not so conspicuous as those of the preceding ; but it would make a pretty border shrub, bearing some resemblance to the Laurestinus, with which it has been compared ; the foliage, however, is very unlike, being of a light-green colour, veiny, and lobed, coarsly-toothed and slightly downy underneath. The fruit is dark purple, or black, hard and flat, not edible. There is a larger species which is known as the Larger Dockmackie or Indian Arrow-wood, *V. dentatum* (L.) The Indians used the long, straight, wand-like branches of this shrub, when seasoned by the smoke of the wigwam, for the shafts of their arrows ; but since they have been able to obtain rifles, the flint arrow-heads have fallen into disuse, and are found no more in the Indian wigwam. This primitive weapon (formidable it must have been) is found only on old battle-fields, or by chance the settler picks up one in turning the soil on his new burnt fallow, wonders at the curious shaped flint, and perhaps brings it home ; but more likely casts it away. It is a type of the uncared-for race, whose forefathers shaped the stone with infinite care and pains.

There is another Viburnum,

SHEEP-BERRY—SWEET-BERRY, *V. Lentago*, (L.)

This species is found in rocky ravines, and on the sides of dry hills. The fruit is sweet and pleasant, and when cooked with the addition of red Currants, forms a very nice preserve, pudding or pie. As the work of settlement goes on, many of our familiar wild shrubs and flowers disappear from their old localities, and in time will be exterminated. Many too that might be introduced into cultivated grounds, and prove floral ornaments in gardens, or useful for kitchen purposes, are doomed to be lost or utterly neglected.

Is there no wealthy botanist, with ample means to do so, who will form a garden on a large scale, and gather together the forest flowers, shrubs and ferns of Canada. It would be a work of great interest.

### BUTTON-BUSH—*Cephalanthus occidentalis*, (L.)

A pretty shrub about five feet high, belonging to the *Rubiaceæ* or Madder family, with light-green, smooth leaves, and round heads of closely set whitish-green flowers. The corolla is tubular, slender ; style thready and protruding beyond the petals. The flowers have a sweet faint perfume. This shrub is chiefly found on the borders of swamps in low thickets. The receptacle remains persistent on the bush in dry round button-like heads, whence its common name. I am not acquainted with any particular qualities possessed by this shrub. It flowers in August.

### POISON IVY— POISON OAK—POISON ELDER—*Rhus Toxicodendron*, (L.)

The Sumac family boasts of two of the most venomous vegetables yet known in Canada, viz., *Rhus venenata* or Poison Sumac, and *Rhus Toxicodendron* or Poison Ivy. The former, *R. venenata* (DC.) is an elegant shrub growing in swamps, with shining, smooth, odd-pinnate leaves, and from ten to fifteen feet high, producing when touched a violent sort of Erysipelas, in some cases fatal in its effects. The leaflets, from seven to thirteen, oval, entire, pointed ; the flowers, small, insignificant, greenish, in loose panicles from the axils of the upper leaves ; berries green, smooth, of the size of peas. This is spoken of as the most deadly of the poisonous Sumacs, but fortunately it is of rare occurrence. The common Poison Ivy, however, is only too frequently met with ; it grows in low ground or on barren rocky islands, among wild herbs and grasses, in open thickets, at the roots of stumps, and will often find its way into our gardens. It may be found in cultivated fields, flourishing on stone heaps—indeed, wherever its roots can find soil to nourish the plant the Poison Ivy may be found. Of its injurious effects on the human body I can speak from experience having witnessed its baneful influence in many instances. Gray, describing its noxious qualities, says : " Poisonous to the touch, even the effluvium in sunshine affecting some persons."

There are various opinions regarding the way in which the virus is communicated, and also in what part of the plant it exists, some persons thinking that actual contact is necessary, others that it is emitted from the leaves when wetted by dews and given out in sunshine : and again it is asserted by some to be the pollen of the flowers floating in the air and resting on the skin, which is the cause ; others again say that the poison is given out in a gaseous vapour at dew-fall. All these suggestions may have some foundation. I am inclined to think that the poisonous qualities of the plant are given out in the heat of the day, when the sun's rays are most powerful, and float freely in the atmosphere, as there are instances of persons being affected in daytime when only passing within some little distance of places where the plant abounded, without coming into actual contact with it in any way.

To some persons the Poison Ivy is perfectly harmless. I, for one, have gathered it for my herbarium in all stages of its growth, without receiving from it the slightest injury, while other members of the family have suffered severly from having been near it, or walking among the shrubs where it was growing. It is during the hot Summer months that most of the cases occur, especially in June and July.

The first symptoms are redness about the eye-lids, ears, and throat, which quickly increase to angry inflamed blotches, rising in blisters, the whole face becoming swollen, so as to produce blindness for several hours or days ; the irritation of the skin is very great. Sometimes the poison extends over the arms, and body, and legs ; fever, headache and even delirium will affect the patient, as in cases of severe Erysipelas. Where the constitution is at all unsound, the effects are worse to overcome, and it is one of the evils induced by the virus that it produces in many cases a chronic disposition to break out, year after year, at the time when the plant is in its most flourishing condition. This has generally taken place in June and July. Some Homeopathists are said to treat the case with doses of *Rhus Toxicodendron*, according to their system ; others again use *Belladona*. Country doctors give alkalies,—soda, ammonia, and cooling medicines. The old settlers apply the succulent juicy leaves and stalks of the Wild Canadian Balsam, *Impatiens fulva*, and other cooling herbs, with thick cream ; but I should think that lime-water, given with milk inwardly, and applied outwardly to the skin, as in burns, might prove a good remedy. Where the disease caused by this poisonous plant is so often met with in country places, the most ready and certain remedies should be made known to the public. Physicians who have had no experience of the disease produced by the Poison Ivy are sometimes at a loss how to treat it successfully.

Every one should be acquainted with the appearance of the Poison Ivy, that it may be avoided when out in the country among weeds and thickets, rocks and waters.

This wicked little plant is not without its attractions to the eye ; it varies in height from about one foot to two, but will climb when meeting with support to ten and fifteen.

I have seen it against a stone building, growing along with the Virginian Creeper, up to the windows of a lofty second story building, no one having discovered the noxious intruder, though very different in foliage from the Creeper. The leaves are three-foliate, thin, of a dull palish green, smooth, but not glossy. The leaflets are broad at the base, indented, hardly deep enough to be called lobed, in some instances only a little waved at the margins, pointed, thickened at the junction of the stem. One of the leaflets is generally larger and more lozenge-shaped than the other two, but they vary a good deal in size and form. Sometimes there is a winged lobe on the larger and outer one. Towards evening the leaves droop downwards, exposing less of the surface to the air and night dews.

The plant spreads by means of the roots, which send up shoots from beneath the surface ; the stem of the plant is woody, thickening at the joints of the leaf-stalks. The flowers appear near the tops of the shoots in little upright panicles ; they are of a pale greenish-white ; the berries ripen in August and are of a dead white, yellow, or dun-coloured. About the time of the ripening of the berries the leaves begin to droop earthward and turn to beautiful tints of orange, varying to brilliant scarlet, which, with the white fruit, has a pretty effect.

The Rhus contains a black dye which is indelible, and which no washing will remove. It is a pity that it cannot be utilized. Professor John Lindley says : " An indelible black dye is produced by the juice extracted from the plant," and adds, " This appears to be a property in common with many plants of this order. The *Stagmaria verniciflua* furnishes the black lac which is used as a varnish in Japan. The resin produced by this tree causes excoriations and blisters on the skin. The Cashew-nut is another member of the order, all which are more or less remarkable as dye woods, or for some medicinal uses, or acridly poisonous."

### STAG-HORN SUMAC—*Rhus typhina*, (L.)

Though belonging to a very poisonous order of plants, our common native Sumac is more noted for its useful than its hurtful qualities. Both the common Dwarf Sumac, *R. glabra* and *R. typhina*, are to be found all through Western Canada, in groves, and on old, neglected clearings, on rocky islets, and by roadsides, the seeds being largely sown by the birds that feed upon the berries.

This is the variety *radicans*.

The foliage of the Sumac is very graceful and highly ornamental to the landscape in the fall of the year, when its long, drooping, pinnate leaves, from nineteen to thirty-one foliate, assume the most glowing tints of orange, scarlet and crimson. The flowers are of two kinds or diœcious, in close, conical, upright heads, terminating the branches. The fruit, small round berries, beset with soft crimson acid hairs, which remain persistent on the receptacle, around which they cluster and give to the tree a strikingly ornamental appearance. These beautiful crimson velvet-like cones continue all through the cold wintry weather, forming a continual feast for the late-going and early coming birds. A bountiful provision for those pensioners on God's providence who " neither sow nor reap, and yet our Heavenly Father feedeth them."

The term Stag-horn, I imagine to be taken, not only from the extended branches, but from the fine brown, downy, covering that clothes the branchlets and stems of the leaves and flower-bearing shoots, resembling the velvety down on the young horns of deer when they first sprout forth.

The wood of the Stag-horn Sumac is of a fine yellow colour, and the chips and bark are used as dye-woods. The bark is used in tanning and the root as a powerful astringent and tonic in intermittent fever, while the acid fruit can be converted into a strong vinegar and is so used, I am told, in New England. I have, however, never seen the fruit of the Sumac made use of in this country for any household purpose.

### Smooth Dwarf Sumac—*R. glaba*, (L.)

This is also widely diffused through Canada. It is a pretty shrub but troublesome, from sending up so many shoots ; it rises from a very low size to ten and twelve feet high. It is very similar to the last, but the foliage is narrower, glaucous-white underneath, the eleven to thirty-one sharply toothed and pointed leaflets are very smooth on the surface and taking brilliant orange and scarlet colours before fading. The stem is also smooth and glaucous, like the leaves. There is another dwarf species, *R. copallina*, (L.) found in rocky soil, the chief characteristic of which consists in the winged margin of the leaf-stalks ; it is a lower and smaller shrub than *R. glabra*.

### Black Alder—Winter-berry—*Ilex verticillata*, (Gray.)

This red-berried shrub belongs to the Holly family ; but we have in Canada no tree which takes the place of the British Hulme or Holly Tree, with its glossy, prickly-armed, evergreen leaves, green bark, and brilliant garniture of scarlet berries.

K

> " It is green in the Winter and gay in the Spring,
> And the old Holly Tree is a beautiful thing."

The Holly among the Romans denoted peace and good will, and possibly for this cause was chosen by the early Christians as symbolical of the peaceable character that should distinguish the followers of the Lord Jesus Christ—the Prince of Peace. The earliest notice of decking the churches and dwelling-houses with Holly, is in the reign of Henry VI, by some pious, but now forgotten writer—a chronicler of old customs—who devoutly lamenting over the disuse of some observances in church matters, consoles himself with the remark that "Our churches and houses are decked with Rosemary, Holly and Ivy, with other goodlye shrubbes that keepe ever green; doubtless to reminde us that the childe then borne was God and man, who shoulde spring uppe as a tender floure to live in oure hartes, and there dwelle for ever more."

Our woody, red-berried Winter-berry is the nearest relation we have to the Holly in Ontario, but it is not prickly, neither is it an evergreen.

The crest of the Strickland family is the Holly Tree, of the Gordons, the Ivy. This custom of heraldic bearings, especially the crest surmounting the coat of arms, is very ancient, and may be referred back to the time when writing was not in use, and formed a sort of pictorial history as to the origin of the family. We find it here among Indian tribes, each tribe, and the members of it, being known by its totem, or heraldic sign. Thus we have the "Eagle Tribe," the "Crane," the "Crow," the "Snake," &c. The figure of bird, beast, tree, or reptile, being the sign adopted by the heads of the tribe, or chiefs, as the sign manual to be appended to any deed or treaty; scratched or figured with pen, charred stick, or knife, whatever is the instrument at hand, the totem is rudely drawn, and is the superscription of the tribe, or their totem.

The individual name is derived from some circumstance independent of the totem of the tribe; whatever object first meets the eye of the child is given as a name. Thus we find "Opechee" (robin), "Omemee" (wild pigeon), "Snowstorm," "Red Cloud," "Westwind," "Murmuring Waters," and other poetical names among the Indians, descriptive of natural objects or events.

The Holly is endeared to us by many interesting associations connected with childhood and youth up to extreme old age.

> It gladdens the cottage, it brightens the hall,
> And the gay Holly-tree is beloved by all ;
> It shadows the altar, it hallows the hearth—
> An emblem of peaceful and innocent mirth.

Spring blossoms are lovely, and Summer flowers gay,
But the chill winds will wither and chase them away ;
While the rude blasts of Autumn and Winter may rave,
In vain round the Holly—the Holly so brave.

Though the brave old English gentleman no longer now is seen,
And customs old have passed away as things that ne'er have been,
Though wassail shout is heard no more, nor Mistletoe we see ;
They've left us still the Holly green, the bonny Holly-tree.

—(An old song by an old lady.)—C. P. T.

There is an old couplet that is common in the North of England about the Holly :—

" O the Oak, and the Ash, and the bonny Holly tree,
They flourish best of all in the North countrie."

The dark green evergreen leaves of the Holly, with their rich garniture of vivid scarlet berries, which remain persistent all through the Winter and far into the Spring, have been so often described or alluded to in print, that they must be well known to all, even in the colonies, and from its use in adorning houses and the churches from Christmas-tide till Candlemas, or the beginning of Lent, the Holly is much thought of and valued, by young and old, in England ; but we miss both the evergreen leaves, and the old associations in our Canadian Holly, and so it is less cared for on that account. The bush—for it never rises in this country to any height—is from eight to ten feet high ; it is mostly found in damp swampy soil or on the banks of streams and beaver meadows, partaking of the habits of the Alder, which it resembles in its love of moisture.

The leaves are ovate, somewhat narrowed at the base, serrate at the edges, thin, and not spiny, rather downy underneath ; the branches and branchlets dark coloured ; flowers greenish, on very short stalks clustered in the axils of the leaves ; the bush stiff and upright ; leaves deciduous ; berries bright red, remaining on the branches through the winter ; much sought for by the Wild Pigeon and Canadian Partridge.

There is another species of the same order known as the

MOUNTAIN HOLLY—*Nemopanthes Canadensis*, (D. C.)

which is found northwards in cold bogs. Early in May, the swamps where this shrub abounds, have a warm reddish-brown hue from the colour of the young leaves, this soon turns to a delicate green, which again changes as it gets mature to a bluish glaucous green, the rose-coloured berries are gracefully borne on long pedicels and are sometimes in great profusion, when they present a beautiful effect. The berries of these hardy shrubs are a great resource for food to the " Wee hopping things," our late and early birds, and together with the dry seeds of the Mullein and Rough Amaranth, which harbour many insects in their husky seed vessels, support them till the Spring returns bringing food and gladness to the earth, when the Great Father opens his hand and filleth all things living with plenteousness.

# THE CANADIAN FOREST.

"Not such thou wert of yore, ere yet the axe
  Had smitten the old woods, their hoary trunks
  Of Oak and Plane, and Maple o'er thee held
  A mighty canopy."—*Cullen Bryant.*

"A glorious sight, if glory dwell below
  Where heaven's magnificence makes all the show."

LTHOUGH the snow lingers longer within the forest than on the open, cleared lands to which the sun and winds have more ready access, yet vegetation makes more rapid advances, when once the Spring commences, within the shelter of the trees.

No chilling, biting, winds or searching frosts penetrate the woods,—to nip the early buds of leaf and flower as on the exposed clearings. Within the forest all is quiet and warmth, when without, the air is cold and the wind blustering. It is among the low bushes and young saplings that the first tints of early Spring verdure are seen; under the kindly nursing of the shrubbery we find the first Spring flowers and succulent plants. The hungry cattle seem instinctively to know that it is in the forest they will find food suited to their wants; leaving the dry fodder that has been their support through the long winter months, we see them hastening to the woods, however deep and miry the way, to browse on the tender, swelling buds of the Sugar-Maple and Basswood, or searching for the oily blades of the Wild Garlic.

Let us go to the forest as soon as the snow has disappeared from the leaf-carpeted ground; we shall see the seedlings of many plants springing up from among the decaying leaves at our feet.

That prostrate plant, with slender stem and pointed leaves arranged so prettily in whorls of fives or sixes, is *Galium triflorum*, sweet-scented like our English Woodruff; and that bright-green, cheerful looking herb, that spreads in creeping mats over the dead leaves, is the pale-flowered *Veronica officinalis.* There are Winter-greens—the Pyrolas, of several species. We find them all fresh and green as when the feathery snow first hid them from our view. The foliage is of a dark shining green, which gives one the idea of endurance against cold frosty weather; near by you may see the graceful fronds of the evergreen Wood Fern, *Aspidium spinulosum*, though bright in colour, yet beaten down and somewhat torn from the weight of the snow that has been pressing upon it ; and there where the soil is more rocky, the dark shining fronds of *A. acrostichoides,* a hardy, handsome fern, known by its smooth scythe-shaped leaflets, and stiff upright growth. The soft parsley-like leaves of the Sweet Cicely, *Osmorrhiza brevistylis*, refresh the eye with their bright verdure ; and as the warmer airs of April are felt, the ground is brightened by the starry blossoms of *Hepatica triloba*, the lovely Snow-flower, that like the English Daisy comes with the first sunny days of Spring ; you may see them in the forest and in the open spaces of groves and thickets, white, blue and pink; and if you wish to transplant them to your garden they will bloom as kindly there as in their native woods. Shelley says :

> "After the slumber of the year,
> The woodland Violets reappear."

Yes, we have Violets of many hues : white, azure, pale blue, lilac, yellow. Some low and delicately small, others larger, more conspicuous, many-branched and tall. There is the pure Canadian Wood Violet, a very lovely species, that in the garden will bloom twice in the season, with graceful branching stems and milk-white flowers ; but these Violets come later in May and June, along with the branching Yellow Violet.

The Yellow Violet, like the White Canadian Branching Violet, loves the leaf-mould and the deep shade of the forest trees. The early White Violet, a small inconspicuous flower, and the Canadian Violet, are the only ones that have any scent, and then it is but a faint perfume.

In damp, mossy soil, see those trailing garlands of Nature's own weaving, the elegant *Linnæa borealis*, with its twin-bells of pale striped pink.

Another of our creeping forest plants is the graceful Partridge-berry (*Mitchella repens*), a lovely fragrant flower with an abundance of small dark glossy leaves and starry white blossoms. Another tiny-leaved little Evergreen plant, mostly found in Cedar swamps, is the Creeping Snow-berry, with trailing branches and white waxen fruit. This is not the shrubby Snow-berry, but a very low Evergreen creeping plant, *Chiogenes hispidula*.

One of the prettiest of our early forest plants is the *Smilacina bifolia*, (Ker.) with small white starry flowers ; it is nearly as sweet-scented as the elegant cultivated Lily of the Valley, which lends its name to our little forest species. Then there are the Claytonias, with delicate pencilled pink flowers ; and, just at the edge of the forest, the pure ivory-petalled Blood-root *(Sanguinaria Canadensis)* opens its starry blossoms to the sunshine on bright Spring days.

Mingled with these fair children of the deep forest shades are Ferns —graceful, elegant Ferns—and Club Mosses, like miniature Pine trees. A kindly nursing mother is the forest, to these her lowly offspring : the earth their cradle, the pure snow their coverlet, warm, soft and light, to shield the tender nurslings from the Winter's cold and biting winds.

Before the shrubs in our gardens have made any show of greenness, in the warmer shelter of the woods, the Fly-Honeysuckle has put forth its bright green leaves, and the soft brown downy winter-buds of the Moose-wood have burst and shown the yellow funnel-shaped clustered flowers. How carefully had these little flowers been protected and guarded from injury on the grey leafless branches through the frosts of winter in their downy coverings. How little do we understand the beneficent nature of that Great Creater who careth even for the embryo leaf and flower.

To those who love the forest and its productions, the continual destruction of our native trees will ever be a source of regret, even while obliged to acknowledge that so it must be, for with the change of soil must necessarily disappear many, or indeed, most of the rare indigenous plants that are sheltered by the woods and nourished by the decaying vegetation of the trees and shrubs beneath which they grow. Exposed to the force of drying winds and hot sunshine, these children of the soil perish and are no more seen.

That close observer and sagacious writer, John Evelyn, in his valuable work on " Forest Trees," writing of the denuding of the forests in Italy and other European countries says : " We find the entire species of some trees totally lost in countries where they once abounded, as if there had never been such planted or growing in them. Be this applied to Fir, Pine, and several other trees; accidents in soil, air, &c., which we daily find, produce strange alterations in our woods. The Beech almost constantly succeeding the Oak, to our great disadvantage."

This author elsewhere deprecates the destruction of the forest trees in England, and the necessity for planting to replace the more valuable timbers—the Oak and Pine. Evelyn wrote and published his " Sylva " during the reign of the last of the Stuart Kings, forseeing the time would come when the country would have to be supplied with her building material from other lands.

Circumstances continually re-produce themselves. May not Evelyn's remarks apply to our Canadian forests ? Especially to the Pines and other Coniferæ, which are being cut down by wholesale in our woods, and converted into lumber.

So rapid has been the consumption of our Pines, that there are townships which have been so stripped of these trees, that in a few more years a full grown Pine will not be seen. As the Pine disappears a change takes place in the atmosphere and in the soil. It is true a new race of vegetables takes possession of the ground, but something has been lost.

The ultimate destruction of our native vegetable productions, including the valuable timber of our forests, which long series of years could not replace, is not the only change that arises from the clearing of a large portion of our woods. There is yet another and important result which will in course of time, be felt as an evil—I refer to the drying up of the inland streams and smaller tributary waters. It needs but little observation and is patent to the older settlers of the Dominion, that the creeks and rivulets which formerly flowed through their lands, are disappearing with the clearing away of the woods. The water-courses are grown up with Sedges and coarse aquatic herbage, and the thirsty cattle now wander far afield in search of water, unless duly supplied by the farmer at the homestead, or driven, at the cost of much time, to springs and water-holes, which are kept open with difficulty during seasons of drought.

In many cases the sources that give rise to the streams might have been preserved fresh, and free from drying up, by allowing a growth of trees and bushes to remain about the head waters of the springs. The existence of springs is generally indicated by small Sedges, Water-ferns, Wild Persecarias, Mimulus, Brook-limes, Arums and Marsh Marigolds, with sundry other water-loving plants " that have their haunts by cool springs and bubbling founts, or by the rushy margin of the stream." The wild animals and birds need no guide to direct them to these secret reservoirs. With no compass to steer by, they are led by an inward power which we call instinct, to spots where their needs will be supplied.

I remember meeting with an old volume in my father's library, and in the quaint language of old Anthony Horneck·were the words, " Doth God take care for oxen?" The answer was brief "Yea, God doth take care!" That was all—but it was sufficient, because borne out by His words who could not err, knowing the mind of His Father: "Consider the ravens," saith Christ, "for they neither sow nor reap, which neither have storehouse nor barn, and God feedeth them."

It seems now to be an established fact that the climate of many countries has been materially affected by the total destruction of its native forests. If this be so, then surely it behooves the legislators of this country to devise laws to protect future generations from similar evils, by preventing the entire destruction of the native trees. There are large tracts of Crown Lands yet in the power of the Government, and reserves might be made or laws enacted by which the valuable products of the soil might be in some measure protected. Let our wise, far-seeing statesmen see to it.

"A tree is a round volume bound in its own bark. Each page, from heart to skin, registers a year of age and growth. The botanist may not only read the record of these leaves, but read the whole constitution of the tree, the laws that govern its vital functions ; may study and understand the system of its veins and arteries, the circulation of its white blood (the sap) and the whole machinery and process of its nutrition and growth. All this is written by the same finger that he recognizes in man's physical system."--*Chap. vii, p. 212 Burritt's "Chips."*

There is a quaint remark made by an old writer, on forest trees, quoted by Evelyn : "Trees and woods have twice saved the world, first by the ARK, then by the CROSS, making full amends for the evil fruit borne by the Tree in Paradise by that which was borne on the Tree in Calvary."—*Evelyn Sylva, Book IV, p. 300.*

THE CANADIAN PINE—WHITE PINE.—*Pinus Strobus*, (L.)

> We paused amid the pines that stood
> The giants of the waste,
> Tortured by storms to shapes as rude ;
> With stems like serpents interlaced.
>
> How calm it was, the silence there
> By such a chain was bound,
> That e'en the busy Woodpecker
> Seemed stiller by the sound.—*Shelley.*

In the brief outline which I propose to give of the native forest trees of Canada, the Pine seems naturally to claim pre-eminence, both on account of its noble growth, and its great value as a source of wealth to the Dominion, whether we regard it from a commercial point of view or as a means for affording employment to a large portion of the industrial classes, especially the *habitans* of Lower Canada. It would require the knowledge of a practical merchant to calculate the value of our Pine forests when summed up in all departments. Some idea may be formed of the importance of this branch of trade by even a casual glance at the vast piles of Pine boards and timbers, laths and shingles that are ready at every port along the St. Lawrence and the great lakes, to freight the vessels that are waiting to bear off the ever accumulating mass to the destined markets—east and west ; to England or the United States. To distant islands and foreign lands, our noble trees, in the form of lumber, find their way. It would be a curious history could we follow one of our grand old forest Pines,—from its first development in the backwoods—a tiny slender thing, of a few thready-spiny leaves—to its towering height and pillar-like grandeur, lifting its dark plumy head above its compeers, drinking in the light and rains of heaven,—to the time when it measures its giant length upon the ground, brought low by the axe of the sturdy chopper. It would be vain to follow out the destiny even of one such mighty Pine, or to weave a romantic history of its voyagings, its wanderings, and its uses. So, leaving the imaginary, we will take up again the sober thread of our subject.

Extensive as is the reign of the Pine tribe in this country of woods and forests, forming a large proportion of the native trees, it has probably at some distant period occupied a still further range than it does now. In the hardwood lands—where the largest Pine trees are now found growing, singly or in isolated groups, from three or four to perhaps a small group—the resinous substance commonly known as *fat pine* is found in larger quantities and in finer quality than that on the pine ridges where the trees are more abundant. This fat pine is the residue of concentrated resinous knots, and roots, where the mighty trunks of which they formed a part have long since crumbled into dust; now Oaks, and Beeches, and Maples, in every stage of growth, from the hoary tree in extreme old age to the tiny seedling, occupy the soil where once those giant Pines grew and flourished. The decay of the Pine is a slow process—more than a century, perhaps two or three, must have passed over before one of the massive trunks, to which those knots and roots belonged, would have become so completely decomposed as to leave no trace behind, excepting these almost imperishable portions. Some of the pieces of fat pine are so saturated with the oils and resinous secretions as to assume somewhat the colour and fragrance of fat amber, an article that is often found in small nodules and water-washed fragments on the beach of the eastern shores of England.

The forced marches of civilization have wrought such wondrous and rapid changes in what used to be the backwoods of Canada forty years ago, that now it seems almost a thing of the past, to write about or to speak of such matters. The writer recalls to mind the old time when in early Spring the waters of the still lake, with its dark Pine-clad shores used to be enlivened with the canoes and skiffs of the fisher, stealing out from the little bays and coves, with the red glare of the fat-pine all ablaze, casting its stream of light upon the dark surface of the waters, from the open-grated iron basket or jack, as it was called, raised at one end of the little vessel on a tall pole. In those days the lakes and inland waters swarmed with fish, which formed one of the resources for the table of the backwoods settler. But now, the saw-mills and saw-logs, the pine-bark and the saw-dust, have driven away the fish by rendering the water unhealthy and poisonous, and the game laws have told hard upon the poor Indian also. The little fishing skiff, and the fish-spear, like the natives, are passing away.

The pine-knots, still however, have their uses in lighting up the caboose fires on the lumber rafts, and, may be, in the far backwoods shanty the settler's wife still performs her evening task of sewing and knitting by the blaze of the pine knots and roots, which the younger children have collected before the wintry snow has hidden them away under its cold, fleecy covering.

There are still lingering among some of the older settlers, those who can recall to mind the time when lamps and candles were hard to obtain, and the evening light was supplied by these homely gleanings from the forest. I have seen a cheerful circle gathered round the wide hearth so lighted up. The litttle ones shared the rugs of the bear and and wolf skin with the favoured hound and shaggy retriever, while the glancing light fell on the swiftly plied knitting-needles of the mother and elder sisters, and the father sat quietly enjoying the cheerful scene, and rest from a day of manly toil, or superintending some rustic work of his sons. Nor was there any want of pleasant talk or memories and tales of better days, to entertain us as we sat listening in that log-house by the light of the pine-knots. Ah, well! if those days of the old pioneers in the backwoods had their privations, they also had their pleasures : they remain as way-marks on the journey of life, and are not without their uses.

The White Pine generally occupies the ridges of light land above the shores of lakes and streams, not flourishing on the low alluvial flats and swampy ground. In wettish soil, such as old beaver meadows, the tree becomes gnarled, and knotty and misshapen, throwing out many rugged, twisted branches, and is utterly useless as timber.

On casting your eye along the border land of any of our inland waters a distinct series of vegetable productions may be noted, each belt distinguished from the other.

First, then, we perceive on the ground nearest to the water, rooted in the deep alluvial soil, dwarf Willows of several kinds, the Red-barked Cornel, Black Alder, American Guelder-Rose, Poplars, and some kinds of Hawthorn ; and wreathing these in leafy-tangled masses, the Frost and Fox Grape vines. Then come Cedars, Black Ash, the fragrant Balsam Poplar and Balsam Fir. These moisture-loving trees fill up the lower range. The stately White Pine towering above takes the high ground, often in a continuous belt, while the deciduous, or hardwood trees, which seem ever pressing onward, take the tableland—a Benjamin's portion—seeming ever bent on encroaching on the pine limits, fulfilling their great mission, that of preparing for man a more fertile soil, better suited for the operations of his hands and the growth of the life-supporting cereals. The decomposition of the leaves, bark, and woody fibre of the Oak, Basswood, Beech, Maple, Cherry, and other deciduous trees, is in God's kind providence a source of fertility, of the blessings of which man is ultimately the recipient. Yet he that receives the gift is often unmindful of the way in which for unnumbered ages it has been preparing for him, by agents appointed for the work. These unconscious labourers have silently been fulfilling the will of Him " who commandeth and it is done."

A noble object is one of our stately forest Pines rising in one uninterrupted column. The grander to the eye as it measures it, for the very simplicity of its outline, and we repeat with the poet :—

" Than a tree—a grander child earth bears not."

Looking upwards, the eye follows its massy shaft rising in solitary majesty—" fit mast for some high admiral ; " and such its probable destiny if chancing to grow in the vicinity of lake or river shore it come within the ken of some adventurous lumberman (your Jean Baptiste has a specially keen eye for a good stick of timber), its fate is sealed.

Soon the lonely echoes of the forest are ringing with the blows of the sturdy axeman on the devoted trunk—and many a vigorous blow is struck before that forest giant inclines its dark-plumed head, and with a rending crash, measures its length upon the groaning and trembling earth.

The height of one of these large Pines, varies from a hundred to a hundred and fifty feet in height, and occasionally reaches a higher altitude. A lumberman told me that he had cut nine saw-logs, each measuring twelve feet in length, from one Pine, besides, leaving the butt end in the ground, four feet high.

Yet even a tree of this size sinks into insignificance when compared with the giants of Oregon and California. The *Wellingtonia gigantea* which reaches the enormous height of two hundred and fifty feet, three hundred, and even nearly four hundred feet ; or the gigantic Araucarias of the ancient world.

The roots of the Pine do not strike so deeply into the ground as might be supposed, but grow more horizontally, almost on the surface. This one circumstance accounts for the frequent sight of upturned trees of great size. The feathery heads of the Pine rise on an average fifty feet above the tops of the tallest hardwood trees. In the rich and generous soil of the Beech and Maple woods, the Pine attains its greatest bulk and height. There, straight, tall, and robust, it looks indeed the monarch of the woods, unequalled even by the stately Oak so often called the King of trees.

When growing in open ground as on some of our plain-lands where the soil is light, the Pine develops an abundance of lateral branches and a bushy head, which give it so different an appearance, that you might be inclined to regard it as a distinct species, quite unlike the Pine of the forest. These branching feathery Pines seldom attain to any great size and are very handsome objects, with their dark evergreen boughs clothing the stem even to the ground, but they are only useful for ornament in the landscape. As timber they are worthless for building purposes.

In the dense forest it is not till it has surmounted the tops of the adjacent trees, which have hitherto disputed its right to a fair share of air and light, that the Pine is able to develop its branches. Up to this period of its life its course has been upwards, always upwards—its branches few and weak and but scantily clothed with leaves, scarcely give promise for its glorious future—it has had to work its way under many difficulties, but having once obtained access to freer air and sunshine, it increases in growth rapidly. The comparative height of the Pines may be seen at a glance by casting your eye along the dark line that divides them from the hardwood trees. They stand in serried ranks, their arms extending on either side in a horizontal direction like an army drawn up in line. Each whorl of branches answers for a year's growth. The usual way in which the age of a tree is ascertained is by counting the rings of wood, each ring counting for a year, but this is not a perfectly accurate method, as in its early infancy these woody deposits cannot be ascertained, and a time may come when the tree, having attained to its perfect maturity, may continue to exist as a tree, long after its vital functions have ceased to add to its yearly substance to any appreciable amount. There is another way in which we may approach to a knowledge of the tree's age, this is by counting the whorls of branches which are added year by year till it has attained its full meridian height. The leaves deepen in colour till about the beginning of July when they have reached their usual size. This growth of leaves endures the intense cold of winter but as the frost intensifies they lose their verdure and acquire a sombre blackish hue. A perceptible change has come over the evergreens, even these hardy natives of the forest seem to mourn the absence of the warm sunbeams, and to be sensible of the iron rigours of a Canadian Winter.

In April the rising of the sap is felt in every branch, fresh energy pervades the tree in every part. A deep refreshing greenness enlivens the dark dull foliage, and the Pine tribe, retouched by the breath of returning Spring, stands forth in renewed beauty long before the bare, leafless trees of the forest have put forth one single green bud. The new growth of the yearly shoots does not take place till the month of May ; it is but the refreshing and retinting of the old leaves that comes to cheer our eyes thus early in the season ; and as we look upon the rich verdure we call to memory those sweet lines of Mrs. Hemans, so familiar and so descriptive of our Pine woods, in the " Voice of Spring : "

> " I have looked on the hills of the stormy North,
>    And the Larch has hung all its tassels forth ;
> The Pine has a fringe of softer green,
>    And the earth looks bright where my steps have been."

The cone of the White Pine appears a little later than the new shoots, but near the top of the wood of the former year ; they are narrow, curved of a deep or rather bluish green, soft and leathery, slightly pointed, and often covered with clear drops of turpentine, which becomes white and hardened in the course of the year. The winged seeds lie at the base of the scales, imbedded in the leathery covering, carefully secured from injury during its embryo state. The ripened seeds form the food of a large number, both of our birds and smaller animals. The seedling pine is a pretty, tiny, tufted thing, with a slender stem, and a number of dark green needle-like leaves. Look at this pigmy, can it be the original form of yonder stately tree ? And yet it is so. Every year a new set of shoots springs from a conical scaly head at the top of the main central stem of the former year's growth. From this head are developed from five to seven straight upright shoots ; of these the middle one is the longest and strongest, and forms the leader ; sometimes accident, as wind or frost, or insects, injures this central shoot, and two of the nearest and stoutest take its place, so that a double crown is formed.

After a little while the scales that had protected the young spiny leaves fall away, leaving the leaves in clusters of fives, clothing the fibrous woody stems of the new growth which hardens as the season advances. The leaves deepen in colour, and by the latter end of June and July the cones begin to form in the older trees.

* The yearly growth of the new Pine shoots measures from eighteen inches to fully two feet, in a healthy free-growing young tree ; but in the dense forest the length of the main shoots is still longer. The bark of the Pine for many years remains smooth and green. As the trunk increases from within, rifts in the surface, near the roots, begin to appear,

---

* The age of a pine tree, till it reaches its meridian height, has been reckoned at a period of from one hundred to one hundred and fifty years. This is as regards its upward growth ; but does not include the full duration of the tree while living.

increasing year after year as the tree comes to maturity. The bark has roughened and divided into rugged masses, deeply channelled somewhat lozenge-wise, becoming of a whitish grey without, but of a deep, brick-red within, lying in thin layers one upon the other. In the Red Pine the bark exfoliates, and is thrown off in shell-like plates in the older trees. In very old Pines, the bark thickens to the depth of some inches. Within this crust various insects deposit their eggs—each trunk containing a world in itself of insect life.

The great Red-headed Woodpecker, with others of the tribe, attack these trees ; instinct teaches them where they may find the hidden food in the greatest abundance.

From early dawn till sunset calls them to their rest, the forest resounds with their noisy labour, tapping, rapping, rending, till large sheets of bark already loosened by the worms beneath, strew the ground in broken fragments, while the tree, naked and bare and desolate, stands among its fellows with death and decay stamped upon its pillar-like trunk. It is a curious sight that stately column all graven as with some curious grooving tool in a thousand fanciful devices—some like a rare intaglio all deeply cut in curved and wavy lines, as by some cunning hand, the tracery varying in length, and depth, and breadth according to the size and nature of the insect labourer. There are some forming the most delicate and elaborate lace patterns, others as if an attempt had been made to imitate the stem and branches of a tree. These things are the work of the borers and sawyers.

The inmates of a new log house or shanty in the bush are often startled by the curious sounds that arise during the still hours of the night, for it is then that they are chiefly noticed, and the wakeful good-wife wonders what can cause the monotonous creaking, rasping, noise that she hears for hours together, or what has made those heaps of fine sawdust lying on the cleanly swept floor below the unbarked walls of her cabin. These sounds and these heaps of sawdust are the work of the indefatigable sawyers enlarging their domiciles within the bark of the Pine logs.

These sawyers are large flat-bodied worms of a yellowish colour, with red heads and strong jaws ; the upper part of the creature's body which is composed of many flexible rings, is broader than the lower. The surface of the body is rough and adheres to the finger when you touch the skin. The creaking sound is produced by the creature gnawing the wood upon which it feeds. These insects are among the countless hosts that make their dwelling in the forest trees and bring them to destruction by slow but certain steps.

"In the Pine forests of the Southern States," says Nuttall, "thousands of acres of trees have been destroyed by insects in their larval state, some not bigger than a grain of rice."

The Woodpeckers, which have borne the charge of destroying the trees in search of these worms, only attack those in which these insidious enemies have already destroyed their vitality. In the bark of the healthy tree the bird finds nothing to repay his labour—let us give the Woodpecker due credit for his sagacity.

" In all labour there is profit," says the wise king, and, depend upon it, the Woodpecker does not spend his hard work for nought.

Though the Pine tribe, with the exception of the Larch, which is deciduous, does not lose the foliage of the spring at the time the hardwood trees cast their leaves, yet they too throw off their leaves, but it is of former years, some say the leaves of three years age—certain it is, that no sooner has the increase of the present year ceased, than a gradual fall of leaves begins to take place and continues silently and imperceptibly all through the summer months. And so on to the Fall, the dead and useless foliage drops to the earth till a deep carpet of the pale, golden, thready leaves is strewed beneath the tree, on which the foot of the passer-by may fall unheard, as if shod with velvet shoes.

How beautiful, how grand are those old Pine woods ! The deep silence that pervades them ! How solemn the soul feels—as if alone with the Great Creator, whose mighty person is shadowed dimly forth in His works ! There is music, too—deep, grand, solemn music— when the wind is abroad, and sweeps the tops of those mighty crested pillars above you ; in softer, lower cadences, it touches those tender harp-strings, or swells with loftier sound in one grand hymn of praise.

It seems as if one could never exhaust the subject, so much might yet be written on the Pines of our own Canadian forests.

But we have only entered upon the subject of the Pines and cone bearing trees, and must proceed to describe our other species.

### RED PINE—*Pinus resinosa*, (Ait).

The Red Pine is distinguished by its handsome foliage, its smooth red bark, exfoliating in thin plates, after the manner of the Plane and Shell-bark Hickory trees. The height of the Red Pine is from fifty to eighty feet, the wood is fine-grained, of much durability, and valued for its uses in architecture. It never reaches the height and size of the White Pine. The cones are hard and woody, often clustered several together, the leaves, which are borne in pairs, are bright green, from five to six inches long. Where the Red Pine abounds the soil is light

PLATE VI.

THE PINK WATER-LILY  (*Nymphæa odorata*, var. *rosea*).

and sandy, or rocky. This tree is generally spoken of, but quite wrongly, as the Norway Pine.

HEMLOCK SPRUCE—*Abies Canadensis*, (Michx).

" The groves are God's own temple."—*Anon.*

" Spring-dressed in tenderest green,—
   There the young Hemlock spreads its fan-like sprays,
   To court the balmy breeze that scarcely lifts
    The lofty Pines that tower above it."

One of the loveliest and most graceful of our forest trees, is a young Hemlock. As great a contrast does that elegant sapling, with its gay, tender green feathery sprays, bear in its beauty of form and colour to the parent tree, with its rugged stiff and unsightly trunk and ragged top, as the young child in its youthful grace and vigour bears to the old and wrinkled grandsire. The foliage of the young Hemlock in the months of June and July, when the Spring shoots have been perfected, is especially beautiful; the tender vivid green of the young shoots, at the end of the flat bending branches of the previous year, appears more lively and refreshing to the eye in contrast to the older, dark, glossy, more sombre foliage, which they serve to brighten and adorn. The Hemlock does not reach the lofty height of the White Pine, though in some situations it becomes a giant in size, with massive trunk and thick, bushy head; the bark is deeply rifted, dark on the outside, but of a deep brick-red within; the branches are flat, the small, oval, soft cones appear later in the summer on the ends of the shoots of the previous season. The timber of the Hemlock is very durable, tough, and somewhat stringy, loose-grained, but is said to resist wet; it is used for granary flooring, rail-ties and some other purposes in out-door work. The bark is used largely in tanning. The backwood settlers stack the Hemlock bark while clearing the forest land and carry it in during the sleighing season to the tanneries, receiving a certain value per cord, in money or store goods; formerly the payment was chiefly made in leather, when every man was his own shoemaker, but times have changed since those early, more primitive days, and the wives and children would now disdain to wear the home-made boots and shoes, that were manufactured out of coarsely-dressed leather by the industrious father of the family, in the long winter evenings, as he worked by the light of the blazing log-fire with his rude tools and wooden pegs.

The old shanty life is a thing of the past; the carding and spinning, the rattle of the looms, even the knitting needles are not now so constantly seen in the hands of the wives and daughters as formerly. Railroads and steamboats, schools, and increase of population, have wrought great changes in the lives and habits of the people. Villages

L

and towns now occupy the spots where only the dark forests of Pine and Hemlock, Maple and Beech once grew. The trees disappear indeed before the axe and fire, from the site where Nature had placed them, but they reappear now as ornaments, planted by the hand of taste in the gardens, and as shade-trees on the streets of the towns and cities ; and this is good, it speaks of taste and culture. The Hemlock, however, is less frequently seen about our dwellings, beautiful as it is, for it is tardy in growth and does not take kindly to cultivation. Its natural soil is dry, rocky, or gravelly land.

A remarkable hoof-like fungus, of a deep red colour, semi-circular in outline and elegantly scalloped at the edges, with curved lines like some large sea shell, hard, dry, and varnished on the surface, is found occassionally growing on the rough bark or big, scaly roots of decaying Hemlocks. These fungi *(Polyporus pinicola)* are found in clusters of larger and smaller growth on thick stems united at the base. I have seen a group of these singular parasites, the largest measuring more than a foot in diameter ; it was greatly prized for its elegant form and rich colour. The under side of the fungus is of a fine warm buff tint.

Nearly allied to the Hemlock is the well known

CANADIAN BALSAM FIR—*Abies balsamea,* (Marsh.)

A tall, slim, graceful tree, is this beautiful evergreen, distinguished by its dark green foliage and spire-like form. Rising among the lighter deciduous trees of the forest, it makes an agreeable figure by force of contrast with the spreading Beech, and the full-leaved Maple and Bass-wood. In wet ground the Balsam Fir runs up tall and slender, forming thick groves of wand-like growth. The timber of this tree is little used, except for rafters in outdoor buildings. It is the smooth, thin bark of this Balsam Fir that yields the "Canadian Balsam" used in medicine. The clear, resinous juice is obtained by incisions made in the bark, from which it is allowed to flow into reservoirs. The tree abounds with this fluid, so much so that the bright, clear, drops may be seen on the green cones, sparkling like dew in the sunshine and filling the air with a pleasant fragrance.

The leaves are flat in single file on each side of the horizontal branchlets, white underneath, the branchlet ends in three slender sprays When covered by fresh fallen snow these drooping sprays give the tree the appearance of being loaded with lovely white flowers—a sight so fair, few that have ever entered the forest after a snow storm can forget ; it is a sight to delight the eye and lift the heart above earthly things to the throne of the great Creator who has made all things so lovely here, even the snow, as an emblem of His wondrous purity and Holiness. I remember the cross-like form of the upper shoots of the Balsam being pointed out to me one day by an old Irish chopper.

"You see Mistress" he said—touching his ragged napless hat as he spoke : "That even in these wild woods the Lord's cross may be seen pointing to the sky above our heads, to remind us of the Blessed Saviour's self who died to redeem our souls—sure it is well for us to have something to remind us of Him in this haythenish place."

When growing free, on open dry ground the Balsam increases in bulk, forming a large sized, regular, pyramid-shaped tree with sweeping branches of dark glossy foliage, attaining a height of from fifty to sixty feet, and upwards. I never see a group of our beautiful Canadian Balsams with their exquisitely symmetrical spire-like forms, rising from a broad base to the slender pointed apex, but they seem to be silently pointing heavenward, to remind one of the great Creator who called them forth and gave them their beauty of form and their enduring verdure.

### BLACK OR DOUBLE SPRUCE—*Abies nigra*, (Poir.)

This species seems to prefer dry open ground and springs up spontaneously on old neglected side lines and waste rocky places. The foliage is sombre in hue, thinner and more wiry than that of the White Spruce ; the spiny leaves are arranged in brush-like form round the rough branchlets. The persistent cones are small, the scales thin, and waved at the edges.

The Black Spruce is used medicinally by the Indians and old settlers in cases of rheumatic pains, as an ingredient in vapour baths, and its buds are used for the preparation of Spruce Beer. It is planted as an ornamental evergreen.* Fine as this tree is it does not equal in beauty the

### WHITE SPRUCE.—*Abies alba*, (Michx.)

A charming object is this beautiful evergreen when not crowded and dwarfed in its free pyramidal growth by the too close proximity of other forest trees. The White Spruce is seen to most advantage at the edges of old concession and side lines or similar cleared places. Where it can have free access to light and air, there it expands its low horizontal branches and sends up its strong shoots forming a fine outline, and a pleasing contrast with its pale glaucous foliage, to the darker Balsams and Black Spruces that surround it. The new shoots are of a delicate

---

\* The timber is also valuable where the tree attains to any size ; in favourable localities it will reach a height of seventy and eighty feet, with a corresponding bulk ; the timber is light, strong and elastic.

light green, and look like tassels appended to the bluish stiff branches of older growth.    The cones are long, from one and a-half to two inches, and slightly curved, appearing pendent on the last years branches. The White Spruce differs from the Black, in that the larger cones instead of remaining persistent on the branches, drop from the tree as soon as the seeds are ripe.  The spiny leaves surround the branchlets on all sides, and are stiff and rigid as they get older, but soft in the earlier stage of growth.

The Spruce firs are rapid in growth and bear transplanting well, especially if planted in groups, and supplied with water when first removed from their forest soil.    By planting several of the young trees near each other, they shelter one another and retain the moisture about their roots which would otherwise evaporate too quickly.

April is preferred by some persons for transplanting evergreens, before the sap has started the new shoots.   Others prefer July or August, when the sap is less active, but with watering and care Spruce will move well at any time during the Spring and Summer months ; when once rooted they make rapid growth.    Hemlock and White Pine take less kindly to change of soil and place.   As an ornamental tree the White Pine is inferior to the Spruce firs unless in extensive grounds where it has ample space allowed for the development of its branches.

The timber of the White Spruce is white and light ; it is not valued so highly as that of the Pine, nor does it attain to so great a height or bulk.   Hurlburt gives it only 50 feet in altitude, and from 12 to 18 inches in diameter.

### Tamarac.—American Larch.—*Larix Americana*, (Michx.)

"And the Larch has hung all its tassels forth."—*Hemans.*

One of the loveliest heralds of the opening Spring is the Larch, delighting the eye, long wearied with watching the tardy unfolding of the leaves of the hardwood trees ; its bright verdant tufts of foliage, bursting from every spray, encircle the rosy, hardly developed, cones, like a tender green thready fringe.   The more we look upon this beautiful tree the more we find to admire in it.   The young twigs are covered with a golden hued knobby bark ; from this rough bark spring the light thready whorls of leaves at short intervals, each enveloping a soft cone of crimson colour—or cluster of staminate flowers—arranged along the pendent branchlets, which hang gracefully downward, looking as if adorned with strings of fresh rosebuds—or tufts of golden threads.  The branches of the Larch grow at right angles with the trunk, and spread horizontally, slightly drooping at the extremities.   The bark is whitish-grey and scaly, not rifted like the White Pine.   It is a more graceful tree in its growth than the Pine, Spruce, or Cedar, and deciduous in habit, like the European species.

The American Larch is also called Black Larch, but for what reason I cannot determine ; and also by the Indian name, Hackmatack.   The yellow, tough, slender rootlets of the Larch are used by the Indians in sewing their birch-bark 'canoes, and are called *Wahtap.*   The outer bark is removed by steeping the roots in water for some time, when it peels off readily.   The roots are rendered supple, smooth, and very pliant, and are made up in coils for use when required ; the *wahtap* thus prepared is as strong as any cord, and is invaluable in the manufacture of the Indian canoe—more suited to the purpose than any manufactured article the native could adopt—easily obtained and without cost.   The Indians make use of the inner bark of the Larch as a poultice in drawing hard obstinate tumours, but it is a strong and painful application.

The timber of the American Larch is much valued, especially in ship-building, and for railway ties.   On dry, hilly land the Larch attains to a much larger and finer growth than on low, wet, swampy soil ; the wood is much better grained and more compact.   While the value of the White Pine is lessened by growing in open spaces, where it can develop its lateral branches, the Larch seems to improve in quality and attains a larger stature and produces more valuable timber.

While the Larch is one of the first of our forest trees to put forth its glad green leaves it is one of the last to shed them, and lingers long with us, brightening the faded woods with its bright golden colours where all around is sad and grey and dull in the landscape from which the glory of the Summer has departed.

In low, wet, spongy flats the Larch grows tall and slender and is little valued, excepting for rafters or such purposes.   Formerly a Tamarac swamp was regarded as utterly worthless, mere waste lands, a harbour for Bears and Wolves ; but as the country becomes denuded of its woods even the despised Tamarac is utilized ; and when opened out in course of time the soil is cultivated as grass lands and runs for cattle. Many of our inland creeks and springs take their rise from, and are cherished in the deep shade of the Cedar and Tamarac swamps. When these reservoirs are cut down and destroyed much of the fertility of the land will be lost.   People are only now beginning to learn the uses and value of the trees that they destroy ; looking only from one· point of view they regard tHem as enemies.   " Cut them down, root and branch, why cumber they the ground ? " is the cry of the backwoodsman But a day comes, when his eyes being opened by education and experi-ence, he begins to plant a shelter around his bare homestead, and no longer wonders that the fresh-flowing stream that was so great a comfort and pleasure, sparkling as it ran through his pastures, dries up and disappears: the trees that sheltered it and the leaves that caught the moisture from the atmosphere are destroyed.

I know extensive farm lands where scarcely a tree has been left, even as a shelter for the cattle during the hot days of Summer. A thriftless thrift this might be called, where even a few acres could not be spared for the supply of the household fires in future years.

THE WHITE CEDAR.—AMERICAN ARBOR VITÆ.—*Thuja occidentalis*, (L.)

Those frequently occurring and often extensive tracts of land called Cedar Swamps form one of the remarkable features of the low-lying lands of the Canadian wilderness : deep tangled thickets, through which the foot of man cannot penetrate without the aid of the axe, or his eye pierce beyond the limits of a few yards, so dense is the mass of vegetation that obstructs his view of the interior. A secure hiding-place for the wild denizens of the forest is the Cedar Swamp. Within its tangled recesses lurk the Bear, the Racoon, the Fox, and when these are absent, the timid doe and her fawn rest secure from the gun of the wary hunter. The wily Indian cannot molest them within these impenetrable solitudes ; and here wild birds of such species as do not migrate to warmer latitudes, retire during the frosts and snows of the Winter season.

It is from the edges of the Cedar Swamp that the first hollow drumming of the Partridge is heard in early spring. The rapid hammering sound of the Woodpecker greets the ear of the axe-man, or the whispering notes of the little Tree Creeper, and the pleasant cry of the little Chickadees, as they tumble and twirl and flirt among the evergreens, chattering to one another, as if rejoicing in the return of sunshine and bright skies once more, and the bestirring of the insect tribes that lurk beneath the sheltering bark of the old White Cedars.

A mass of fallen trees, deep beds of mosses, rank swamp grasses, and sedges, ferns, and low bushes, and seedling evergreens, occupy the spongy, porous soil, and conceal the stagnant water that lies fermenting at their roots in those dismal swamps.

Silent as the grave, and damp and lonely as they appear, life— insect life, swarms here. Let us pause for a few minutes to examine that huge trunk that lies athwart its fellows, bleaching in the snows and rains of many seasons. It looks sound, but strike it with your axe, and you find it is a hollow cylinder ; beneath the white and gray shreddy bark the woody substance is perforated into countless cells and intricate labyrinthine galleries, the mysteries of which we strive in vain to trace out. The plan to our eyes seems all confusion. Doubtless if we could view the architecture with Ants' eyes, we should perceive—

> " Disorder, order unperceived by thee ;
> All chance, direction which thou canst not see."

These long galleries and cells are the work of a large black Ant. These ants are somewhat formidable-looking insects, of a reddish-black colour, about half an inch in length. The female, or winged insect, is the largest ; then the male ; the workers are of smaller size. There are myriads of these last in that old Cedar, and in those prostrate trunks that lean in every direction above it. These black Ants are among the most active of our forest scavengers ; ever busy, boring, sawing, pounding and tearing ; manufacturing a walled city out of the fragments of those fallen trees ; silently and secretly do they carry on their labours, like the sappers and miners of a besieged city.

A troublesome colony of black Ants is sometimes introduced into the log-house of the backwoods settler in the foundation logs, which are very frequently made of Cedar, being more durable than any of the hard-wood timbers. These creatures soon find out the housewife's stores of maple-sugar, and molasses, and preserves, and carry off quantities, to say nothing of what they devour individually ; and very difficult it is to dislodge and destroy these depredators. They seem to be omnivorous, nothing eatable coming amiss to them. I have seen them stop their homeward march to devour crumbs, dead flies, and even make a meal off the body of their comrades. I remember being greatly molested by a colony of black Ants, when living in a log-house, our first residence in the backwoods. They formed two regular bands, one going the other coming to my store closet. I killed them by hundreds ; but the black brigands never seemed to diminish, till at last I found out their strong-hold, which was a large Cedar post to which my garden gate was hung. This was perforated all through by the labours of these insects, and being close to the walls of the house, they made their entry between the logs ; boiling water, applied in sufficient quantities, at last ridded the house of the nuisance.

The timber of the White Cedar is very light and durable, and is valued above all other for the sills of log buildings, for rafters and posts and rails. The Cedar Swamp, which in the early days of the colony was looked upon by the settler as a useless waste of land and a loss, has now become a valuable possession—in many situations a profitable one. In some places a thousand Cedar rails will realize from twenty to thirty dollars, and even more than that sum in parts of the country where rail and fencing timber is scarce ; which owing to the improvidence of many of the older class of settlers, is now a common case. I have known Cedar rails cost thirty dollars per thousand ; and the buyer had not only to pay this high price, but to cut down and draw home the logs a distance of seven and even ten miles.

When cleared, these Cedar Swamps make good meadow land—the stumps and roots are easily burned or pulled out—and after a series of years, if well drained, will produce root and grain crops, and good pasturage for cattle.

At one time it was a common practice with farmers, when making fences about the homesteads, to reverse the Cedar posts, inserting the upper end in the ground and the butt end uppermost, under the impression that by so doing the wood was preserved from decaying. I think the practice was objectionable, as, from the spiral growth of this tree, the heaviest end of the post was uppermost, forming a lever, which had the effect of heaving the fence out of the ground, when the soil was softened by the action of the frost and thawing rains in the Spring of the year ; besides, the fences so constructed had an unsightly appearance. If the preservation of the wood were the object in view, charring the end of the post before inserting it in the post-hole would have been a far more certain method of ensuring them from decay,

Gray gives the average height of the White Cedar as from 20 to 50 feet, but it sometime exceeds that height. The stem is tall, straight and tapering upwards to a narrow point. Instead of forming a branching or bushy head, the branches curve downward, being wider and more sweeping towards the lower part of the trunk ; very often they are re-curved toward the extremities. The leaves are closely appressed, or imbricated, lapping over each other in four rows on the sharply, two-edged branchlets, which are flat and horizontally placed    The scales of the cones are soft and blunt ; the seeds winged all round ; the flowers are of two kinds, borne on different branchlets.    The Greek name for the Cedar is derived from some resinous tree—possibly from the Cypress, to which it bears a near affinity.    In its early growth, within the shelter of the forest or by the banks of lakes and creeks, the bark of the young Cedar is smooth, and of a dark shining green ; but where it grows in open, exposed ground, it is hard, rough, and scaly, and of a greyish colour ; the foliage is also of a lighter, more yellowish tint of green than in the forest.    When the tree attains to maturity, the bark splits into long lozenge-like divisions, and peels off in ragged strips. The long sweeping branches become rough and hoary in age, in the crevices of which the grey tree-moss (*Usnea*) fixes its long pendulous tufts, and gives to the tree that venerable aspect that has obtained for it the name of White Cedar, in conjunction with the whiteness of the wood and outer bark.

When dry, the wood of the White Cedar is highly inflammable, burning with great rapidity, and leaving only a residue of fine white ashes which are said to be deficient in the fertilizing salts of the deciduous or hardwood trees.

The gum of the Cedar is clear and colourless, and possesses a fine aromatic scent, which is given out after showers and during sunshine. The Indians regard the gum of this tree as possessing very healing and medicinal qualities. They chew it as a pleasant luxury; it excites the flow of saliva, and no doubt is far less injurious to the system than tobacco, of which they are so fond. The Indians use the root of the Cedar as well as that of the Tamerack (or *wah-tap*) in making their Birch-bark canoes, and of the inner bark the squaws weave mats and baskets. The fibrous bark is soft, pliable, and tough, and is better adapted for the manufacturing of mats than that of the Bass-wood. The thwarts of the Birch canoes are also made of split Cedar. The tough silvery-grey paper which forms the outer covering of the wasps' nests is chiefly derived from the fibrous portions of the White Cedar. While watching these industrious insects tearing off fragments of the silky-thready bark from some old fallen tree, I have thought that a manufacture of paper. or felt, might be produced from this abundant material, for which it seems particularly adapted. The fibre is white, shining, and tough; it can be beaten to any degree of fineness; and moreover, seems to be of a more enduring substance than hemp or flax, as I have known portions of Cedar bark to lie on the ground for a very long time, trodden down by the foot, and exposed to every vicissitude of weather, and yet retain their qualities unchanged. It would not be the first time that man has profited by the example of the lower animals in his manufactures, or borrowed from them materials for his work. Who will try to improve upon the paper made by the despised wasp?

Among its many uses, the Cedar has of late years been adopted for garden hedges. It is easily obtained; takes root readily; is extremely neat and ornamental when trimmed with the garden shears; is evergreen, and does not intrude upon the borders, as it sends up no shoots from the roots; is close, warm, and sheltering. Whether it would be proof against the weight of cattle pushing through it I cannot say, unless planted within rails or pickets, as is usually done in gardens. The Hemlock also makes a very pretty garden fence, and possibly the White Spruce, if headed in, might be rendered equally if not more serviceable for enclosures. A very handsome evergreen fence of mixed trees of the above named species would be very ornamental and more serviceable than the Hawthorn, which, in our native species, is hard to cultivate, having a tendency to grow too high and straggling to make a close, compact fence.

On dry soil the White Cedar forms pyramidal groups growing close and compact from the ground upwards, the horizontal branches being so closely interwoven as to appear like one dense bush, and presenting a fine mass of rich evergreen foliage during the greater part of the year, though the severe frosts of Winter change the bright verdant hue to a sickly yellowish tint; but, like the Pines, the rising of the sap early in April renews all its bright colour and reclothes it with fresh beauty. The fall of the old leaves takes place in the latter end of summer, soon after the new shoots and fruit have been perfected.

Pursh, our oldest Canadian botanist, writing of the American White Cedar (or *Arbor vitæ*), says: "Its geographical range is from the northern parts of Canada to the mountains of Virginia and Carolina;" but adds, that "in the Southern States it is becoming rare, and is now only found native on the steep rocky banks of mountain torrents."

The White Cedar takes a more northerly range than the Cypress (*Cupressus thyoides*), also called White Cedar. This latter species prefers a warmer climate, extending southward, whilst our native Cedar is seldom found south of the Alleghanies. Thus each species maintains its own special boundary, retreating by almost imperceptible degrees, and giving place to its advancing rival.

It is not often that the Cedar is found growing promiscuously in the forest among hardwood trees, and rarely, if there, does it attain to any considerable size. When a group of these trees is so found, they indicate the presence of springs; often the head waters or sources of forest streams are thus made known to the exploring woodsman and hunter. The Pagan worshippers of ancient times would have deified the moisture-loving Cedars, making them the sylvan home of Naiads who had their haunts by cool stream and shady grot, or by the rushy margin of lonely springs and bubbling founts.

But though the Cedar is mostly found growing on the low-lying margins of lakes and rivers, yet it is a singular fact that it is frequently found forming dense masses, in detached groups, on high, dry, gravelly ground, and grassy wastes that have long lain unoccupied save by weeds and poverty-grass. In such unlikely spots these Cedar bushes take root, never growing up into tall trees as in the moister lands, yet spreading continually till they effectually cover the ground, and, by excluding the sun and wind, convert, in process of time, the soil into a damp one, no evaporation taking place from the surface through the dense mass of branches that cover the earth even to the very roots of the bushes. The snow that falls in Winter, and the rain in Autumn and Spring, saturate the ground with a superabundance of moisture, which ascends not again in mist or dew. Rank sedges and other moisture-loving herbs, and mosses, and fungi, take the place of a more healthy vegetation. A change is effected, both in the soil and its products, which might lead us to the conclusion that many of our Cedar swamps have thus been originated where once a very different order of things existed. Such facts are suggestive of the changes that are continually taking place in the country, and are not without interest to students of the physical geography of our land.

"This Cedar is the most durable wood in Canada," says Professor Kalm, a writer of the last century, in his travels in North America who enters largely into the uses and valuable properties of the White Cedar. He writes thus : " The enclosures are made of the White Cedar, the palisades of the forts, the planks, the houses, and the thin narrow pieces which form the ribs of the bark canoes used here are of Cedar, being when fresh cut, pliable and easily bent, and very light. The branches of the Cedar are made into besoms by the Indians, and everywhere used in the houses, and give out a peculiar and pleasant scent." The writer then gives directions for making use of the Cedar medicinally in cases of rheumatism : " The green leaves being well pounded are boiled down in lard and applied as a plaster, which eases the violence of the pains very shortly. An Iroquois Indian told me that a decoction of Cedar was used by his people as a remedy for coughs, colds and consumption. This acted probably as a sudorific, also used in intermittent fevers."—*Kalm.*

Evelyn speaks of the salubrious nature of resinous trees, among which he mentions especially the Pine, Juniper, Firs and Cedars. He suggests the introduction of Cypress and Cedar wood for sanative purposes into our dwellings in the form of wainscots, shelves, tables, and other articles of household use, which he quaintly observes " would greatly cure the malignancy of the air."

It was the custom, possibly adopted from this writer's influence, at the time of the Plague in 1666 to burn large bundles and boughs of all sorts of resinous woods in the streets of London, to purify the infected air. Many carried sprigs of Cedar or Juniper in their vests or hands.

I have been told by lumbermen that it is not the choppers and hewers and other men that work in the Pinewoods that suffer from agues and intermittent fevers, the resinous exhalations of the Pines being invigorating, pure and healthy. It is the settlers on newly cleared hardwood land where the virgin soil is opened to the influence of the rays of the sun, and it is the gases that are set free from the moist vegetable mould that are so injurious to health.

### COMMON JUNIPER—*Juniperus communis*, (L.)

On rocky islands and gravelly banks of lakes and rivers, appears to be the natural habitat of the Juniper.    It is a hardy rugged shrub, for it can hardly be said to arrive at the dignity of a tree.    The branches spread horizontally on the ground but become ascending close and bushy above.    The bark is rough and scaly, the foliage thin and spiny, of a dark sad green in the older growth, but the more slender pointed sprays that are liberally put forth in Spring are of a light and tender green.    The flowers clustered on the older branches are greenish, followed by small round berries covered with a white bloom, the ripened fruit later in the Summer bluish-purple.    The whole bush gives out a strong and peculiar scent, especially after a hot day at dew-fall. Like all trees that affect a rocky soil, the roots are strong, tough and wiry.    It is found growing spontaneously far to the North in Canada, but becoming more dwarfed in the colder regions it spreads closer to the ground, covering the sterile rocks with its spreading branches, affording a warm shelter to the wild birds and small quadrupeds that have their dwelling in such inhospitable spots, where plants of a tender nature cannot thrive.    The abundant berries of this hardy shrub supply a table in the wilderness to feed

> " The wild flock that never need a fold."

The bountiful Father forgets not to feed even the lowliest of His creatures, "these wait all upon Thee that thou mayest give them their meat in due season."    These denizens of the lonely Northern regions, that have their haunts among far off lakes and rocky islands, are equally cared for as those that dwell in more sunny and fruitful lands.    He apportions their food to suit their several necessities—there is something for all—even the hard dry juiceless berries of the Juniper are utilized.

> " Nothing useless is or low,
>     Each thing in its place is best ;
> And what seems but empty show,
>     Strengthens and confirms the rest."

### RED CEDAR—SAVIN—*Juniperus Virginiana*, (L.)

This hardy evergreen is found native in Canada, growing on bleak, dry, rocky, hilly banks, and bare sterile islands, where its tough, wiry roots insinuate themselves between the crevices, and thus firmly anchored it will stand the chilling wintry blasts.

But, nevertheless, in these bleak exposures it becomes oftentimes a stunted and rugged looking tree of small dimensions and scanty foliage without beauty of form, and with little value for any purpose. Whereas, in better soil and under more favourable auspices, the Red Cedar is a slender but graceful tree, attaining to a height of from twenty to thirty feet, and the timber from eighteen inches to twenty-four in diameter, but not often exceeding this measurement. The foliage is of a rich, deep green on the older branches; but lighter and of more vivid colour on the slender rapidly-growing shoots in Spring; the leaves are pointed, spreading, in pairs or threes, closely pressed, scale-like, overlying each other. The wood is much valued for ornamental inlaying and cabinet work, on account of its fine pink colour and peculiar grain, and fragrant agreeable scent. But these rare qualities are very evanescent, both colour and sweet smell are lessened by exposure to light. The wood takes the application of French Polish well.

Even the gnarled fantastic roots of the Red Cedar can be made into useful and ornamental articles by being cut and polished—walking sticks from the knotted branches—rustic hall chairs, and garden seats, from the twisted roots; and even shawl pins and brooches have been fashioned from the resinous knots, which appear when polished of a fine semi-transparent ruby colour.

These are only a few of the uses that can be made of the wood and roots of the Red Cedar, but there are others that are well-known to the carpenter, such as panelling for rooms and wardrobe shelves; where the tree grows in abundance, fencing rails and articles of household use are also made from the wood, such as tubs, pails and boxes.

The Red Cedar repays cultivation, becoming a very handsome ornamental tree, thriving well in open ground and generous soil. It is found both Northward and Southward, becoming more rugged and dwarfed in barren localities, and colder and higher latitudes.

AMERICAN YEW—GROUND HEMLOCK.—*Taxus Canadensis*, (Willd) or *Taxus baccata*, (L.) *var Canadensis*, (Gray).

This pretty little evergreen is found chiefly on gravelly banks, in forest land near lakes and rivers, and on rocky islands; it is of low stature rarely rising above two or three feet from the ground which it covers with its low prostrate branches. The leaves of the Ground Hemlock are of a bright shining green, lighter on the under side, two ranked and flat on the branch. The fertile flowers green, solitary and scaly at the base, forming as they mature a little hollow waxy pulpy cup of a beautiful rosy-red; within the hollow of this lovely gem-like cup lies imbedded the dark green seed. The beautiful fruit of this pretty evergreen, so tempting to the eye like delicate pink coral or porcelain, is sweet and mucilaginous; but the central green seed is poisonous. The procumbent branches put forth fibres from the under side, rooting in the soil, thus widely extending the plant over the ground. It never forms an upright stem or woody trunk, like the Yew of Europe.

The Ground Hemlock is the only representative of the Yew Family, in Eastern Canada. In British Columbia *T. brevifolia*, closely resembles the European *T. baccata*, and forms a small tree. In England, in the olden time the Yew was valued for its elastic wood and tough roots, of which the Archers made their bows. Though no longer planted for the manufacture of this warlike instrument, there are still many noble specimens of the Yew to be seen in England, in old parks, and in country church yards. As the funereal Cypress marks, the resting places of the dead in the East, so the Yew still may be seen in lonely country church-yards in England, venerable trees sacred in the eyes of the inhabitants as associated with the dead of ages gone by.

There is (or fifty years ago there was, to be seen) a very magnificent old Yew tree growing on an open pasture near an old red-brick house, not far from Cheshunt Park, once the residence of Oliver Cromwell, or of one of his sons. A grand lofty spreading tree of picturesque form. This noble tree was supposed to have existed on the spot, where it grew, from the time of the early Plantagenets, it was in its meridian glory during the reigns of the Tudors and Stuarts, and was still green and vigorous shewing no signs of decay in the year 1831. The noble tree might even at that late date be said to be enjoying a green old age.

The foliage of the European Yew is a full bright green, lighter on the under side ; the beautiful rosy wax-like cup, in which the purplish green seed lies sheltered, distinguishes the Yew from all others of the Coniferæ.

## The Oak Family.

"Not a prince in all that proud old world beyond the deep,
  E'er wore his crown as loftily as he
Wears the green coronal of leaves,
  With which Thy hand hath graced him."—*Bryant.*

### White Oak.—*Quercus alba*, (L.)

When found growing upon open ground where it has space enough to expand its stout limbs, and form a free rounded and spreading top, the White Oak presents a noble appearance, not inferior in picturesque beauty to the far-famed English Oak ; this species will grow on lighter land than suits some other hardwood trees. Whether the wood of the Canadian White Oak is equal in grain and durability to the English Oak, I must leave to the shipwright aud builder to determine.

I know that vast quantities in the form of lumber, and hewn timber, yearly leave the Canadian ports for consumption in the Old World.

The White Oak is widely diffused throughout Canada, the average height in the forest being about 130 feet, but on open plains it seldom grows so high, but throws out wider spreading branches. The diameter of a fine Oak trunk is from 60 to 84 inches. In the forests of the western peninsula, the White Oak attains to its largest size. The wood is much used as staves for casks, spokes, naves of waggon wheels, railway ties, beams, &c. The bark is largely used in tanning, the weight of a cubic foot of the wood, when fully seasoned, is fifty pounds. The ash of the White Oak contains a good amount of potash, but is not equal to that obtained from Hickory, this last-named tree is the best for burning, and for the production of potash of any of our native trees.

The acorn of the White Oak is ovate or egg-shaped, very smooth, of a bright reddish-brown, sweet, but astringent. The leaves stiff in texture, oblong, with from five to nine rounded lobes. When young the leaves are very white, or reddish-white, and downy, especially in the dwarf variety, the White Scrub Oak of the plains, or Oak openings as these open grassy tracts of land are commonly called, many acres of which are densely covered with a thick growth of dwarf Oaks of several kinds, grey, white and black. These dwarf varieties spring from thick, knotted root stocks, which send up several sprouts, or woody stems which never reach to any great size, but are used by the settlers for firewood and poles for fencing.

Beside being extensively used in tanning, the bark of the White Oak is employed medicinally as an astringent. It is also used in dyeing. The thrifty wives of the country farmers make a black dye for wool and cloth with certain proportions of oak bark, copperas and log-wood. The inner bark is chiefly used for such purposes. The squaws manufacture strong coarse baskets for farm purposes from the inner bark of the Oak, which parts easily into strips after being rendered pliable by soaking in water and pounding with a heavy wooden mallet.

The old English poet, Chaucer, has some descriptive lines on an Oak grove which are curious, and point out a fact that I have myself taken notice of in my forest walks, *i.e.*, the regular distances that trees, in their native uncultivated state, occupy in the forest when planted by Nature's own cunning hand. The quaint and obsolete spelling of the words will amuse :—

> " A pleasant grove
> " In which were Okis grete—strait as a line,
> " Under which the grass so fresh of hue
> " Was newly sprung ; and an *eight* foot, or *nine*,

" Every tree well fro his fellow grew.
" With branches brode—laden with levis new
" That sprengen oute agen the son-shene ;
" Some very rede, and some a gladsome greene."

*—From Chaucer's poem of " The Flower and the Leaj."*

The regular distances with which the hardwood trees, such as Beech, Maple and Basswood, grow in the forest from each other, has often been a matter of speculation ; from eight to ten feet appears to be the usual distance where the trees have attained to any size. You rarely see the hardwood trees growing in dense thickets ; in the case of the scrub Oaks, and small scrub Pines, they often do grow in dense masses, but in such case they never attain to any size and merely form an underbrush. Among our native Oaks,

BLACK OAK.—*Quercus coccinea*, (Wang.) var. *tinctoria* (Gray.)

Is one of the most useful and valuable of our native forest trees. In height it is found from 100 to 130 feet, and from 4 to 6 feet in diameter. A grand and lofty tree is the Black Oak. The bark, when young, is of a dark green colour and perfectly smooth, but as the tree increases in size and age it becomes deeply divided into large, flatish, oblong figures, and in old trees the bark is deeply channelled and very thick and dark coloured. The inner bark is the *quercitron*, of the dyers, and is also used in tanning. From the dark colour of the bark and deep holly-green of the foliage, the common name of Black Oak has been given to this species. The wood has a handsome but somewhat coarse grain, and becomes of a deep tint when long exposed to the air and when in household use. The leaves of the Black Oak are of a splendid shining dark green, deeply lobed, and sharply pointed. The acorn globular, or nearly so, harsh and bitter, the cup covering about half the acorn, is rough and scaly.

The head of this noble tree is more compact and bushy than the White Oak, the limbs not so large nor widely spreading from the trunk. This fine species grows to great size on open loamy soil. It is not so picturesques in form as the type of the species, the

SCARLET OAK.—*Quercus coccinea*, (Wang.)

This is an exceedingly beautiful and ornamental tree. The foliage is of a rich dark glossy green, which turns to a brilliant scarlet in the Fall when touched by the frost. The lobes are deeply cut, and divergent from five to seven in number, smooth and shining. When growing in open spaces, the tree spreads and obtains a fine and handsome form.

The leaves of the Scarlet Oak are set on long slender stalks ; the acorns are roundish, somewhat deep in the cup, and not so large as those of the Black Oak.

### THE RED OAK.—*Quercus rubra*, (L.)

This is a fine large spreading tree, from seventy to one hundred feet in height, with a diameter of from three to four feet where it grows freely and in suitable soil. It blossoms in May and bears large acorns in very shallow, smooth cups (or rather saucers). The foliage is handsome, with wide-spreading lobes ; the wood is porous and coarse-grained ; the leaves turn to a dull red when touched by the frost and remain persistent till the Spring, when they are displaced by the swelling of the new buds. The wood is chiefly used by coopers for casks, especially for oil and molasses casks.

The acorns of the Red Oak are eagerly eaten by hogs, and the smaller wild animals lay up stores of them in their underground granaries. The wood burns well when it is well dried ; the weight of a cubic foot is forty pounds.

Dr. Lee says that acorns, roasted as a substitute for coffee, have been given to young children affected with rickets and scrofulous diseases with beneficial results. The brown inner coating of the acorn is also valuable and highly astringent in the form of washes and gargles for the mouth and throat, especially in relaxations of the uvula and tonsils—a simple remedy, and easily procured. It is a pity that it is so little known.

### MOSSY-CUP OR OVER-CUP OAK.—*Quercus macrocarpa*, (Michx.)

This is a fine large Oak, with handsome deeply lobed foliage and fringed cups, that grows on the Rice Lake Plains. The cup is large, coarsely-scaled and woody, with fringed border, by which mark it is easily distinguished from any of the other Oaks. The leaves are deeply lobed, almost pinnate, long and slender and of a bright green, the bark, grey and scaly ; a fine whitish fringe of awned scales surmounts the edge of the cup.

### SCRUB OAKS.

The Black Scrub Oak is one of the handsomest of our Dwarf or Scrub Oaks, with dark shining holly green leaves, the lobes are finely bristle-pointed, in height varying from ten to twenty feet, but where it springs from the acorn and grows singly it will make a good sized tree of handsome outline, but this is rare ; the Black and Grey Scrub Oaks

M

send up several shoots from a thick spreading scaly root-stock or stool forming dense woody thickets. The Black Scrub Oak is of little value, even for firewood, and here it is curious to observe that in all the Oaks, those with dark foliage and bark, are of inferior value for fuel and produce little · potash. The housewives know this by experience ; the green unseasoned Black Oak makes bad ashes for soap-making purposes and is wretched burning wood unless thoroughly seasoned, while the grey-barked Oaks burn well and yield more lye from the ashes.

The Scrub Oaks are only found on comparatively light soil and form one of the difficulties of clearing plain lands, as the roots possess much vitality and the shoots must be cut down twice in the season for two or three years successively. After the land has been chopped and the brush burned—at the end of the third year—the exhausted stock ceases to send up more sprouts ; and as they have no tap root and are very superficially attached to the soil, these " Nigger-heads," as they are commonly called, easily yield to the ploughshare and handspike, and are then collected in heaps on the ground and burned, or brought home and stacked in the wood-yard for fuel ; they make a hot fire heaped on the hearth, but are awkward and unwieldy for stove use.

The farmers of the plains do not wait to till the soil and sow their grain till the Oak roots are removed, but make such temporary clearings as the rough condition of the ground admits of, and are content with small returns for the first two or three years ; but after that they reap good crops of splendid grain, and soon have the satisfaction of seeing their fields free from the unsightly stumps that remain so long an eye-sore in the forest clearings ; the fields then present an Old World aspect to the eye.

The Grey Scrub Oak occupies the same localities with the Black variety, yet, varies in bark, in foliage and size. It has been said that the dwarfing of these Oaks arises from the fact that many years ago the Indian hunters made a practice of burning over tracts of these plain-lands, to promote the growth of the various grasses on which the Deer fed, and by dwarfing the young growth of trees made better covert for the wild game that sheltered there among the hills and ravines; but this I think is doubtful.

It would be difficult, now, to trace the fact from experience, for so rapidly has the work of clearing the land and the consequent exterminating of the underwood taken place, that in a very few years, not a specimen of the native Scrub Oak will be found on the Rice Lake Plains. For upwards of fifty years we know that no such process of burning the ground over by the Indians has taken place, and there are large timber trees here and there mingled with the Oak brush. These large trees must be centuries old, judging by their size and ancient appearance ; surely they would have suffered from fire as well as the other trees. For when once the bark of a tree is scorched by fire it ceases to live, and the fire running over the brush and long grasses would have destroyed the vitality of most of the larger Oaks. The last burning of this tract of land must have taken place some hundreds of years ago to have given time for the growth of the timber trees still existing. The Mohawks and Ojibeways had many battles on these plains, the former claiming the hunting grounds on the Southern shore of the lake. Rice Lake is still called in their language "The Lake of the Burning Plains." The ground is now all under cultivation. The Oaks, like the Indian race that hunted over these beautiful hills and vales, are disappearing.

The Indians of the village of Hiawatha, on the North shore of the lake, near the mouth of the Otonabee, can give no distinct account of the time when the hunters burnt the Plains over—it was not in their time—"Mohawks burnt not us, Mohawks bad Indians." This was all the answer I could obtain. One thing is remarkable, that you rarely find any fruit on the Black and Grey Scrub Oaks, and yet the ground is so thicky set with these bushes that it is with difficulty a path can be forced between the stems.

Though I have only enumerated five species of Oak besides the Scrub Oaks, there are several other varieties known to the lumberers in the backwoods. Pursh names no less than thirty-four species of American Oaks, probably many of these are natives on this side of the Great Lakes. The Oak is one of the most valuable of our native forest trees ; it will grow on lighter land than the Beech, or Maple ; it is widely scattered through the country. Even as far North as Hudson's Bay the "Grey Oak" is found, according to Michaux, if Pursh quotes him correctly. Cowper, the English poet, in alluding to the longevity of the Oak, says—

"Lord of the woods, the long enduring Oak."

The Oak was considered to be an emblem of strength and vigour. Mention is made of the Oaks of Bashan in the Old Testament in many places.

We read in chapter xxxv of Genesis, that Jacob, when he came again to Bethel, commanded his household and people that were with him to put away their strange gods, or idols, and he (Jacob) "hid them under the Oak that was by Shechem." A little further on follows the death of Rachel's nurse Deborah ; " and she was buried beneath Bethe under an Oak, and the name of it was called *Allon-Bachuth*, or the Oak of Weeping."

The worshippers of Baal raised their altars under groves of Oak trees ; and the ancient priests of Britain, the Druids, worshipped under the Oak, which was held sacred by them on that account.

The " Oak of Palestine " is an Ilex.   The writer has a small spray sent from the Holy-land taken from an ancient tree in Hebron, traditionally said to be Abraham's Oak.   The tree beneath which the Patriarch spoke with the Lord and the destroying Angels, and where stood the tent in which Sara prepared the feast for the sacred trio.  "If not the tree so celebrated," wrote the reverend gentleman who sent the relic, "No doubt one of very great antiquity."

The leaves of the Oak were used for the civic crown, with which Roman citizens were honoured.

The Oak seems to be more subject to the effects of lightning than any other of the forest trees.  The cause possibly may exist in the mineral matter that enters into the formation of this hardwood tree, or it may be that the Oak is often left to stand as an ornament by itself, in exposed places, open pastures, or parks, and is therefore more exposed to the influence of the elements in times of storm and tempest.

There is a couplet in a poem by Sir Henry Wotton, that has much significance on this subject—

> " I would be high, but see the lofty Oak,
> Most subject to the rending thunder's stroke. "

The Oak-tree is the crest of the Holy-Oak family, and of the Camerons.

In some of the country villages in the eastern counties of England in my younger days, the houses of the villagers were adorned with Oak boughs and leaves on the twenty-ninth of May, the anniversary of the restoration of Charles the Second ; while gilded Oak-apples (galls) were worn by the men and boys as emblems of the Oak of Boscobel, in which the royal fugitive was concealed after the loss of the battle of Worcester.*

THE WEEPING WHITE ELM.—*Ulmus Americana*, (L.)

> " Under the shady roof
> Of branching Elms star-proof ;
> Follow me."—*Milton.*

> " With what free growth the Plane and Elm,
> Fling their huge arms across my way.—*Bryant.*

The White Elm is one of the tallest of the hardwood trees of our Canadian forests, where it attains to a great height.   Straight as a lance

---

* NOTE.—On my grandmother's side my father was descended both from the loyal Cotterel family, and also from that of the Pendrils, of no less noble principles, who withstood all bribes to betray their prescribed defeated Sovereign.   Many were the tales that were told at our fireside, of deeds of loyal faithfulness, that are unrecorded in history, to which my sisters and myself listened

it rears its pillar-like stem, often reaching sixty or seventy feet free of the forking branches ; in diameter frequently reaching four feet. Of all our forest trees few surpass it in grandeur of size, and none in elegance of shape. Free of the forest where it grows singly, and with room to develop its branches, the Elm presents one of the most charming features in our rural landscape, assuming a variety of picturesque forms, not unfrequently reminding one of the grotesque capital letters which are now often adopted to ornament the first words in our illuminated books—an old fashion borrowed from the ancient monkish manuscripts.

The branches of the White Elm divide at the crown of the trunk, and rise almost of uniform thickness to form a level top. From the outer boughs hang down slender leafy branchlets, which, like long loosened tresses, wave with every motion of the wind in the most graceful manner imaginable.

Sometimes the whole trunk is clothed with leafy sprays, which give to the tree the appearance of being clothed with some elegant climbing plant. I have seen these light green sprays mingled with the rich dark green foliage of the Virginian Creeper, forming a beautiful mass of light and shade. In the Fall the magnificent crimson hues of the Creeper touched by the frosts, harmonize in charming contrast with the fading tints of the Elm, producing an effect of rare beauty.

The wood of the White Elm is extremely tough and hard to hew, and makes very poor fuel ; it will lie for years undecayed. In former years, before the value of the forest trees was as well known as it is now, the choppers considered the presence of many of these great Elms a sad nuisance in clearing the fallows, as the wood is hard to burn, and the trunks are hard to cut into lengths for logging—but now the Elm is valued and used for many purposes, especially by the wheel-wright, and the timber is exported to Europe for such purposes as require great toughness and strength.

Beside the White or Swamp Elm, we have two other kinds—the Corky-barked or Rock Elm, and the Red Elm. The latter is better known in Canada by its common name, Slippery Elm, which expresses its qualities and is therefore well adapted to it.

### THE SLIPPERY OR RED ELM—*Ulmus fulva*, (Michx.)

Is possessed of valuable medicinal qualities in addition to its uses as a timber tree. The inner bark and twigs are mucilaginous, healing and

---

with eager ears, and which doubtless had their influence in after years on the minds of two of those enthusiastic auditors. Yet, strange to say, our father, though descended from the loyal Cotterels and Pendils, and the no less loyal Stricklands, was no Jacobite, but a great admirer of William the Third. It might be that his house had suffered from its adherence to the ungrateful Stuart Kings, Charles II and James II, but children are not always influenced in their historical prejudices by the opinions of their parents, as it proved in this case.

softening ; ground up into a coarse powder, it is used in the form of poultices, and a drink is concocted from the inner bark which is given in fevers, and in complaints of the throat and chest ; it is by no means unpalatable, and is nourishing and wholesome.   Our medical men know its value, and often recommend its use.   The Indian chews the young leaves and applies them as a healing application for wounds.

The foliage of the Slippery Elm is oblong pointed and doubly serrate, having an agreeable scent when dry.   The bark on the tender branchlets is downy, as also are the cases of the leaf-buds which are clothed with soft rusty brown hairs.   The tree is not one of the largest or loftiest of our forest trees never attaining the height and dimensions of the gigantic White Elm.   The wood is reddish in colour, whence the name Red Elm, tough and valued for many purposes.   No less useful is the

ROCK ELM.—*Ulmus racemosa*, (Thomas).

A larger tree than the Red Elm.   Of this tree Dr. Hurlburt says : "The Rock Elm is found in most parts of Canada, and grows very large in the western counties, averaging 150 feet in height and 80 to the first limb, with a diameter of 22 inches.   It is abundant in the western part of Canada ; preferred even to White Ash by some carriage and waggon makers, for the poles and shafts of carriages and sleighs.

The wood of the Rock Elm bears the driving of bolts and nails, better than any other timber, and is exceedingly durable when continuously wet ; it is therefore much used for the keels of vessels, water-works, piles, pumps, boards for coffins and all wet foundations requiring wood.   On account of its toughness it is selected for naves of wheels, shells for tackle-blocks, and sometimes for the gunwales of ships.

The bush settlers know its uses in making ox bows and axe handles, it being greatly valued for such purposes.

The branches of this tree are ridgy and winged and the bark is deeply furrowed.   I have noticed that the Rock Elm in this part of Canada grows chiefly on waste poorish soil, and the tree is dwarfish, and has not a free vigorous look.

The European hedge-row Elm is sometimes seen near the roadsides. How introduced one cannot say, but the trees may be known by the small leaves and picturesque growth—always beautiful whether growing by the roadside, on open grassy glades, or copse-wood ; there is not a more charming object in a rural landscape than an English hedge-row Elm.

There are (or were, for I am not sure that they still remain) several remarkable Elm trees in the suburbs of Toronto, that well deserve the special notice of the lover of forest trees.  One in particular, a grand, lofty, spreading tree, near the Orphan's Home.

A tree of such perfect symmetry of outline that it could hardly fail to strike the eye of the beholder with admiration.  It was probably a vigorous tree in its meridian beauty before the first log-house was raised upon the site of Little York.  Now the first city in Western Canada.

What tales could that mute witness tell of toil and privation, among the hardy adventurous few that cleared the forest land on which this now solitary giant of the lonely wilderness stands.  What strife, political and physical, has it beheld.

Beneath its leafy canopy the Red man reared his wigwam, or the early missionary from far off France held up the Cross and preached the word to the listening stolid Indian.  It may be that the seed fell upon the rock and brought forth no increase.

The Indian treads the busy streets, where once the Deer stole forth from its leafy covert in the dense Cedar swamp, where now stands the lofty church or busy mart.  That lofty tree alone remains a memorial of the Indian's hunting grounds, and he himself stalks along those crowded streets the shadow of a dying race.

How many over-wrought minds and toil-worn bodies have sunk into dust ; while still serenely grand and majestic, untouched by decay, the noble tree stands a good Watchman at his post, looking over the rising city ; and still may it stand through storm, and wind, and heat, and cold, a more glorious object than all those stately buildings over which it casts its evening shadow.

BLUE BEECH—AMERICAN HORNBEAM—*Carpinus Americana*, (Michx.)

This tree, commonly known to the backwoodsmen as Blue Beech, is also termed Ironwood by some writers, though it differs greatly in appearance from the Hop Hornbeam or Ironwood proper.   The bark is smoother, more like that of the common Beech ; the wood is whiter and can easily be divided into long slips, and is much used by the settlers and Indians in the manufacture of axe handles and common brooms : these are made by boys, of an evening by the fireside, and in weather that prevents them from out-door work.  A clasp-knife and a straight stick of Blue Beech is all the stock in trade, required for the producing of one of these homely useful articles.   To effect this the bark is removed and long strips are cnt from the wood and drawn down to within a few inches of the end of the stick ; these are then turned over, bound and trimmed even at the ends, and the shanty broom is made, supplying at the cost of an hour or two's labour, a good serviceable substitute for a more costly article. It is by such simple expedients that savings are made in the Canadian backwoods—or it was so in the early days of the Colony.

The Blue Beech is a tree of small size, rarely exceeding 20 feet, often not more than 10 or 15. The bark is smooth and grey, most like that of the Beech, and the foliage resembles that of the Birch. The little nutlets are borne on large foliaceous bractlets, not so large nor so showy as in the

### IRONWOOD—HOP HORNBEAM—LEVERWOOD—*Ostrya Virginica*, (Willd.)

This is a well known forest tree, valuable for many uses. The wood is hard, with a metallic lustre, fine grained and very heavy. The tree is never very large, seldom more than forty to fifty feet in height, and from eight to twelve inches in diameter. The bark is divided into long straight lines. The wood is used for handspikes, levers, reaches of waggons, and rafters, and burns well as firewood. When found growing in open cleared places and copses, it forms a more bushy and shrubby tree, very ornamental through the Summer from the elegant white involucres of the Hop-like seed clusters; the seed itself is hard, nut-like and bony. These Hop-like appendages hanging from every spray, have a very pretty effect. The tree is easily raised from seed, and might be introduced into shrubberies and groves with good effect. In its native woods the Ironwood is found scattered among the other deciduous trees, where it grows straight upwards, making but small head. The foliage is oblong-taper-pointed, sharply and doubly serrate, buds acute, involucre sacs hairy at the base, becoming much inflated by the month of August, when they become brown and shrivelled as the seed hardens and ripens.

### AMERICAN BEECH—*Fagus ferruginea*, (Ait.)

"Where feathering down the turf to meet
Their shadowing arms the Beeches spread."

Of the Beech we occasionally see two varieties mentioned in books and hear them spoken of as the "White Beech" and the "Red Beech."

We have, however, only one species of *Fagus* in Canada. The Blue Beech, commonly so called, not being properly a Beech. To the

---

NOTE—Professor Lindley, in his interesting work "Natural System of Botany," places the Ironwood (*Ostrya*) in the same natural order as the Oak, Beech and Chestnut, while some of the older writers class it with the Birch, under the generic name of *Betula*. The generic name of the Blue Beech—*Carpinus*—is derived from the Celtic words *car* (wood) and *pinda* (head), in allusion to the fitness of the wood for making head-gear, or yokes for oxen. It is curious to trace the meaning of some of the classical names given to plants, which were as familiar to the ancients in former times, possibly two thousand years ago, as are the simple names by which we now call them. Some, indeed, of the old Greek or Saxon names, were very fanciful, and had reference to heathen deities or some strange idea or fancied resemblance to things, the likeness to which we cannot now see.

Canadian farmer, the Beech is indicative of the best Wheat growing soil. The decomposition of the leaves, wood and roots of this valuable forest tree, giving a rich black mould, highly favourable for all agricultural productions of roots and grain, which yield abundantly in the soil of the Beech woods ; fortunate is the settler whose lands consist chiefly of timber of this kind.

Even in its native woods the Beech is a tree of slow growth, throwing out many horizontal branches. Its slow, upward growth, tends to promote the hardness and close texture of the wood, and to give a more symmetrical form to the general outline of the tree. While yet very young, the leaves remain persistent on the boughs through Autumn storms and Winter snows. The light, fawn coloured faded leaves when contrasted with the dark evergreens, with the snow beneath and the blue cloudless sky above, give an agreeable brightness and cheerfulness to the forest scenery that robs the dark lonely wood-paths of much of their gloom.

Growing singly or at the outer edge of the forest, the Beech forms a tree of great beauty from the wide sweep of the branches, at first curving downwards, and then slightly upwards, its slender sprays and shining bright foliage, in early Spring of a tender green, but darkening as the Summer advances, give a lively appearance to the woods. It is a common practice to head in the young trees when planted as shade trees, but this treatment alters the natural, graceful, pyramidical form, giving a rounded figure which becomes thick and bushy, and far less graceful to the eye. The Beech, when in fruit in the months of August and September, acquires a russet hue from the brown, rough, urn-shaped involucre, that contains the three-sided nut, which every boy knows is eatable, and children seek the fallen fruit among the leaves that strew the ground.

Though three-seeded, there are rarely more than two nuts in one husk, one being empty out of the three. The swine feed largely upon the Beech-mast, and become fat upon the oily food ; but nut-fed pork is not regarded with much favour by the housewife, as it is soft and runs more to oily fat than good, sound, hard bacon. The timber of the Beech is used chiefly for such purposes as require smoothness and hardness, it being very compact and strong. In England the wood is much used for turning, and also for flooring ; in this country it is valued for many purposes ; for the handles of Carpenters' tools, shoe lasts, cogs of wheels, and for common bedsteads, mouldings, picture-frames, and many household things, beside turner's wares. Hurlburt says the White Beech averages 110 feet in height, and 50 feet to the first limb of a well grown forest tree, 18 inches in diameter ; but I think this exceeds the average height. As firewood, there are few woods better than good dry Beech ; the green wood is slow to burn, but when thoroughly alight gives out a strong heat ; it is good economy to lay in a stock of green wood in the sleighing season for the next year's use.

If the rough, fallen timbers and the under-brush are removed, the Beech will form a charming home park, beneath the shade of which the natural grasses spring up, and soon make a fine sward, a delightful, cool, retreat for the cattle and sheep during the heat of Summer.

The bark of the Beech is smooth and of a bluish-grey tint, richly variegated in age by the various minute Lichens of many colours that cloud its surface. These give a picturesque aspect to the trunk and larger limbs that afford studies for the pencil of the artist. An old fantastic gnarled Beech is, to the painter's eye, a thing of beauty and of value ; he would, in the enthusiasm of his art, be inclined to cry out

" Woodman, spare that tree."*

### THE BIRCH.

" The fragrant Birch above him hung
Her tassels in the sky."

Of all the deciduous trees, the Birch appears to be the hardiest. It is found in the coldest climates, in far Northern regions, where even the hardy Pine and Spruce will scarcely grow, excepting in a dwarfed and stunted state.

In the far Nor'-West, near Hudson's Bay, the Birch, in a low dwarf form, still may be seen ; it is found in Iceland and Lapland within the limits of the Arctic Zone. Its uses are as numerous as its geographical range is extensive and varied. The Reindeer feeds upon the leaves, which also form a bed for his master ; the twigs and fibrous roots are woven into baskets and Birch brooms are well known in the stables and outbuildings of the farm. The sap, in Russia, is fermented for wine and vinegar ; the aromatic oil of the Red Birch gives the peculiar scent so pleasant to the Russia leather of the book-binder. Who, in Canada and the American States, is not familiar with the bark canoe of the Indian, the embroidered baskets, boxes and mats of the squaw ?

Nature is never idle. No sooner has the flower and fruit of a tree arrived at perfection than a new work commences, nay, in many cases has commenced. New powers are called into action, new leaf-buds are forming and pushing off the old effete foliage, new material is being collected to form the flowers and fruit for the ensuing year. Close-

---

* The classical name of the BEECH is from the Greek word which means " to eat," in allusion to the esculent nuts.

hidden lie the embryo blossoms enclosed in cases, sealed and impervious to the action of wind, tempest, rains, and biting frosts, that might otherwise injure the precious treasures concealed within their warm, protecting bosoms. Some of these leaf and flower buds are covered with a varnished, odorous gum as in the Balsam Poplar ; others lined with soft, silky hairs as the long, taper buds of the Beech ; some with brown wool, as the Elm, the Moose-wood, the Willows, the silky catkins of which are seen in the Winter, and many others.

Take a Small bunch of the common Red-berried Elder, and open the round knobby buds, that appear soon after the leaf has fallen, and lo! there within the leafy cradle that encloses them, lie closely packed, the numerous greenish-white flowers, waiting for the warm breath of May and June to call them forth and expand the cymes of perfect, closely-packed, but undeveloped, blossoms, to the sun-light and the Summer breeze.

Are not such things worth looking for ? Should they not fill our hearts with wonder, love, and praise to Him whose infinite wisdom has ordered all things rightly ; who careth for the creatures he has called into life, yea, even for the humblest herb and the grass of the field.

In some trees and shrubs the swelling of the new buds is not so apparent as in those that I have mentioned, but still the process is going slowly and secretly on, even when hidden from observation.

### PAPER OR CANOE BIRCH.—*Betula papyracea*, (Ait).

> "Where weeps the Birch of silver bark
> With long dishevelled hair."
> *   *   *   *   *   *   *   *   *   *
> "Where the light Birch its loose tresses is waving."

The catkins of the Birch are formed almost as early as the leaves fall. The little, hard, scaly, close-pressed catkins, in threes, appear at the ends of the branches, and may be seen all through the winter months increasing and swelling as the ascent of the sap is felt, till in the months of April and May the yellowish blossoms appear and the long waving tassels flutter in the wind, giving an air of lightness and grace to the slender branches, and making the Birch one of the most graceful and attractive of all the trees of the forest. The Birch, with its snow-white bark and branches, when seen among the Pines and Evergreens, forms a delightful contrast with its airy lighter foliage, to their dark sombre colour and stiff outlines.

The White Birch springs spontaneously on neglected clearings and waste lands ; it is hardy, easy of culture, and truly a great ornament at all times and seasons of the year.

To the North American Indian the Birch is scarcely less valuable than the Palm to the natives of the Tropics. From the tough, pliable bark the Indians manufacture vessels of every conceivable size and shape, from the rude mokowk, or rice and sugar basket, to the most delicate and richly ornamented work-box, pocket-book, and sheath for knife or scissors, curiously worked with moose--hair and porcupine quills ; these materials, humble and apparently of little worth, are dyed with the brightest of colours, and serve the purpose of giving a gorgeous effect to the equally simple, natural article, the bark, of which they are constructed. It is of the large sheets of the White Birch that the Indians form their canoes, and the sides of their winter wigwams are panelled with the same flexible material. These panels are so contrived as to form temporary storing places that answer the purpose of closets and wardrobes ; bags and boxes in which all sorts of miscellaneous household articles are kept, clothing, dried meat, fish, rice or sugar, in their bark baskets, or packs of peltries ; in short, anything and everything is stowed away out of sight in these primitive pockets. Necessity is the mother of invention, and teaches the simple children of the forest expedients at which the white man is inclined to smile, if not to scoff.

The construction of the Birch-bark canoes—those light and portable vessels with which the Indian navigates the lakes and inland streams— is usually the work of the women, who are as skilful in the art of boat building as the men. The shape, length and breadth of the canoe is marked out by upright sharp sticks driven into the ground, thus forming the outline and supporting the sheets of bark during the process of building. The frame work (or ribs) is composed of split Cedar ; the sheets of bark are sewed together with the roots of the Tamarack (American Larch) ; these roots are steeped in water, peeled and coiled ready for use, and look like thick cord—smooth and very tough. Over the seams of the Birch-bark a strong adhesive pitch, made from the resinous gum of some of the Pine tribe, is plastered, and a strong coat of varnish, also made of some resinous gum, is painted over the whole. The thwarts are firmly secured to the edge of the canoe, sewed strongly with the wah-tap, and when all is gummed and varnished the little vessel is ready for the launch.

The dry bark of the Birch is very inflammable being supplied with a fine aromatic oil which is readily ignited, and gives out a delightful aroma while it is burning. The dry bark is sometimes used instead of fat pine in the fisherman's skiff when night spearing on the lakes. Formerly, before lucifer matches were in use the very thinly peeled bark of the Birch was much used as a ready substitute for tinder by the settlers in the Backwoods, and in kindling a fire a bit of this bark makes it burn up rapidly. I am not aware if the bark of the Birch has ever been used as a substitute for white rags in the manufacture of paper, but in its native condition it can be made use of for writing upon either with pen and ink or pencil.

The sap is rarely used here as it is in some of the Northern countries of Europe for making wine and vinegar. Our settlers in the Backwoods prefer the more generous juice of the Sugar maple for that purpose.

The extensive clearing away of the forest is a cause of great regret to the Indians who have to go far back to obtain a supply of the canoe bark. We no longer see the light canoe dancing on the waters of our lakes, as formerly. I remember noticing a Squaw watching the burning of a large log-heap, and in answer to some remark made, she observed with a moody glance at the burning pile : " No canoe now—White mans burn up Birch-tree. Go Buckhorn-a-lake for bark. Got-a-none here," and wrapping her arms in the folds of her blanket she turned away sullenly from the destroying fire, and no doubt there was anger in her heart against the settler, for the trees of the forest which were being wasted by axe and fire. Who could blame her? Now, indeed, the want of the Birch-bark is a sore loss and privation to these poor people. They cannot go into the woods to cut down a single tree, without being liable to fine for trespassing, and the game laws press hard upon them. The Indian race is fast fading away like many of the native plants, we shall seek them in their old haunts, but shall not find even a trace of them left.

Another of our native Birches is the

YELLOW OR SILVER-BARKED BIRCH.—*Betula lutea*, (Michx. f.)

The Yellow Birch grows to the height of 60 to 80 feet ; the bark is greyish with a silvery lustre, but in age the exfoliation of the bark gives a rough ragged look to the tree. The wood is hard, capable of a fine polish, is of a warm yellow, inclining to reddish colour, and is much used for bedroom furniture. The bark gives out a brilliant gaseous flame with a pleasant aromatic scent. The timber of this tree is excellent firewood, and next to the Black or Cherry Birch, is highly valued in cabinet work.

BLACK BIRCH OR CHERRY BIRCH.—*Betula lenta*, (L.)

is the most valuable of the Canadian Birches. It has been known by the term American Mahogany, on account of its colour and grain, and not being liable to warp it is prized by the cabinet-maker among our first-class native woods. The tree grows to a considerable size ; the bark is reddish and smooth when young, speckled and bronze-coloured on the twigs. The leaves of the Black Birch are heart-shaped, pointed and finely doubly-serrate. The catkins thicker than in the White or Canoe Birch. This tree may be found in rich, moist woods, and the leaves and twigs are aromatic, and pleasantly-scented when crushed. Pursh enumerates seven species within the geographical limits of Canada, as known to him, but it is possible other species exist in our now extended Dominion.

AMERICAN MOUNTAIN ASH.—ROWAN-TREE.—*Pyrus Americana,* (DC.)

> " Thy leaves were aye the first in Spring,
>   Thy flowers the Simmer's pride ;
> There was no sic a bonnie tree
>   In a' the countrie's side" O Rowan-tree
>
> " The Mountain Ash
>   No eye can overlook, when mid a grove
> Of yet unfaded trees she lifts her head,
>   Decked with autumnal berries that outshine
> Spring's richest blossoms."—*Wordsworth.*

A more ornamental tree can hardly be seen in our shrubberies than the bright-berried Mountain Ash, the hardy Rowan-tree of Scotland. Our native tree differs from the European species by its slenderer form, the deeper green, and narrower more pointed leaflets, and smaller size of the fruit.

The Canadian species is found growing in the rocky townships north of Peterborough, among the low flats and ravines of the granite ridge that intersects that portion of the country where it is found, forming scattered thickets, conspicuous by the flat cymes of white flowers in the month of June, and by the bunches of brilliant scarlet berries in the Autumn months. The berries are eagerly sought by birds that still linger in our woods and thickets—the Robins, Golden-winged Flickers, and Blue Jays feast upon the ripe fruit, some of the smaller quadrupeds of our forests gather up the seeds. These Autumn birds, however, generally leave sufficient for those charming visitors which come to us during our cold, severe weather in January and February ;—the Pine Grosbeaks and the Bohemian Wax-wings, and even then a gleaning remains for the early Spring birds that come before the buds have opened on bush and tree.

Closely related to the Mountain Ash are the Apple and Pear, the Juneberry and Hawthorn. The common name of the Mountain Ash has probably been derived from the pinnate leaves, which are placed in pairs of from thirteen to fifteen along the mid-rib. In height the native tree seldom exceeds fifteen to twenty feet, but in its favourite haunts in rocky valleys, it is often a mere shrub, the stem being only a few inches in diameter, it can hardly be said to rise to the dignity of a tree in its growth.

The Highland peasants know the Mountain Ash only by the more familiar name of Rowan-tree, to them it seems endeared by some tender home associations, bringing back to them the remembrance of mountain streams and lonely valleys; nor have their poets been silent in their songs to the Rowan-tree. Who has not known and heard that sweet, simple lyric of the Baroness Nairne :—

"O Rowan-tree, O Rowan-tree."

Whose eyes did not fill as they listened to words so touching, because so natural, for it is such songs that take us back to the home scenes of our childhood, and awaken memories of the long, long, by-gone years, when young hearts were as gay, and hopes as bright, as the berries that they strung for necklaces, and felt as proud of their ornaments as a court lady of her diamonds and her rubies.

In Ireland the Mountain Ash is called the Bour-tree.

WHITE ASH.—*Fraxinus Americana*, (L.)

The largest and most valuable of our native trees of the Ash family, common to our forests, is the White Ash, which usually is a tall stately tree of regular and upright growth, well known by its light grey bark, deeply divided into lozenge-like sections, and pinnated foliage, smooth and shining leaflets from five to fifteen; flowers in clustered racemes with keyed, winged, fruit. The wood is very white and light when dry, splitting readily. It is much used for the handles of agricultural tools, such as hay-rakes, hoes, forks, &c. The wood is much valued for its toughness and elasticity and is in great request by the waggon and carriage makers, for the spokes of wheels and shafts. The White Ash is a capital burning wood, easily ignited, and giving out a bright flame and good heat. In its native woods it attains to one hundred feet; it is often found from twenty-six to thirty-six inches in diameter : the size varying very little from the root to the first branches. While the White Ash is chiefly confined to the forest, being rarely seen in any open situation, the Black Ash— *Fraxinus sambucifolia*, (Lam.)—is found on river banks and seems rather to affect a damp soil, wet woods and swamps ; it is of smoother, darker bark and foliage than the White Ash. The younger saplings are used for hoops, and the inner bark for basket work. The knots and roots of the Black Ash are beautifully veined and grained, and are used for inlaying and other ornamental works. Besides the White and Black Ash, the

Red Ash—*Fraxinus pubescens*, (Lam.)—is a common species, easily distinguished by its downy, buds, young leaves, and petioles. It is a smaller tree, and is found in similar localities with the White Ash.

### BLACK WALNUT.—*Juglans nigra*, (L.)

The Walnut family has several noted representatives in Western Canada, all useful and ornamental, but none so truly valuable as the Black Walnut. This noble tree is confined to the Western peninsula, being rarely found in its native state East of Toronto.

The beautifully grained and coloured wood of this tree has obtained for it a world-wide celebrity since the Industrial Exhibition in 1851, where it attracted great attention among the connoisseurs in fine woods, from the rich colours, the feathery-waved figures and violet-tinted shading, and the fine polish of the surface. It is peculiarly adapted for the manufacture of massive dining-room furniture, for side-boards, dining tables, book cases and such articles.

Our Canadian upholsterers lose sight of the fact that the rich heavy Black Walnut wood is not so well adapted to drawing-room furniture, as the lighter Curled, and Birds-eye Maple, which is more suitable where lightness and elegance are required.

In its general character the American Black Walnut closely resembles the European species ; its wide-spreading branches, abundant foliage, and stately trunk render this tree one of the greatest ornaments of Western Canada. The pinnate leaves consist of from fifteen to twenty-one leaflets, which are ovate, pointed, serrated, of a fine, bitter, aromatic scent when slightly bruised. The sterile flowers form long drooping catkins of a rich olive green. The fertile flowers, solitary or in clusters at the ends of the branches, appear in June. The bark of the young trees is much used in dyeing, as also is the root-bark ; the nuts are rugged ; the kernels sweet, but not equal in flavour and richness to the European varieties of the Walnut family.

The heart-wood has a beautiful violet tinge which deepens and turns to various shades of brown, almost to black, after long exposure to the air. It takes a high polish, and forms our most valuable wood for cabinet work. The Black Walnut grows abundantly on the rich soils of the Western and South-western parts of the Dominion between the great lakes. Its average height is 120 feet, 70 feet to the first limbs, and from 3 to 4 feet in diameter ; sections of the wood of six feet in diameter are not uncommon ; it is a fine burning wood, but to use it thus seems a great waste of valuable material.

PLATE VII.

The name of the Genus is derived from the Greek words *Jovis glans*, or the Nut of Jupiter. The next in importance of this valuable tribe is the

## BUTTERNUT.—*Juglans cinerea*, (L.)

Like the Black Walnut, this is a handsome, spreading tree, though not generally so tall. It takes a wider range to the Eastward and Northward, growing freely in open lands, in colder and more exposed situations, fruiting abundantly. All these trees are great feeders, and indicate good soil; but this species may also be found occasionally, on poor, rocky soil sending its large roots to a distance wherever it can find nourishment for the support of the large trunk and wide-spreading branches, which require a great deal of space, light and air for their development. In height the Butternut seldom exceeds 80 to 100 feet, and from 24 to 30 inches in diameter. The wood is valuable, but not so highly prized as the Black Walnut; it is used in panelling. for furniture, and carved work. The Indians make butter troughs, kneading troughs, spoons, ladles, and such small household articles from the wood. The bark is grey and rugged, and deeply furrowed when old. The foliage, like that of the Walnut, is aromatic, and pointed; but of a lighter green in colour; the young bark is clammy, as also the buds, while a clear, gummy liquor exudes from the fleshy green husk that contains the nut. The shell of the Butternut is divided into long ridges; the kernel is dry, sweet and edible, but not so nice as the fruit of the Hickory. An excellent warm, brown dye for wool and woollen goods is extracted from the bark; the process is simple, merely steeping the bark in water till the colouring matter is extracted is sufficient; the material to be dyed is then immersed and left to absorb the dye; the goods should be moved occasionally. This is one of the staple household dyes of the country people, who use it for yarns, stockings and home-spun cloth. The Butternut also possesses some valuable medicinal qualities.

As a shade tree on lawns or in meadows, it forms a handsome object. The young nuts are also used for Pickles, in the same way as those of the Walnut. The wood of the Butter-nut is hard but brittle—it makes good fire-wood, but is now considered too valuable for other purposes to be put to such a use. Another species is frequently found in damp woods known as

## THE BITTER-NUT HICKORY.—*Carya amara*, (Nutt.)

Known by the bitterness of the nut and darker wood.

The leaflets are from seven to eleven, dark-green, smooth, serrate at the edges; the husk of the roundish, thin-shelled fruit, is somewhat

N

depressed ; the kernel intensely bitter.   It is a handsome tree ; but the timber is inferior to that of the other Hickories.  The bark of this species is also used in dyeing.  The Bitter-nut affects a wetter soil than the more popular

WHITE HICKORY.—SHELL-BARK HICKORY.—*Carya alba,* (Nutt.)

The fruit of this well-known species is much sought for when ripe, in the month of October.   It is round, hard, and of a light colour ; the kernel, sweet and pleasant, though wanting the higher flavour of the Walnut.   This tree, like the Plane, sheds its bark, which peels off in large, flat plates, giving an unsightly and ragged look to the trunk and larger limbs.   The foliage is smooth and glossy, of a full, rich green and fragrant when crushed, giving out a fine aromatic scent.

The wood of the Hickory is hard, and is considered to be the heaviest of all our Canadian timbers, not even excepting that of the Oak ; a cubic foot, according to Hurlburt, weighing in pounds 58 ; that of the White Oak when fully seasoned 50 ; the Black Walnut only 30 ; the Sugar Maple 38 : that of the Ash 40 ; the White Elm 36 ; the White Birch 32 ; Black Birch 46 ; Black Cherry 34 ; while the Bass-wood is one of the lowest, being only 26 when dry ; the Poplars and Willows range yet lower in the scale ; while the Ironwood again reaches to 47, and the Tall Dogwood *( Cornus florida)* to 50.  *These particulars may not be uninteresting to some of the mechanics and the dealers in the native woods of Canada.   The wood of the White Hickory is highly esteemed as fuel, and the fruit is an article of ready sale in the stores and fruit markets.

The wood is much esteemed for its elasticity and toughness, being used in many manufactures where these qualities are required.

The bark yields a yellow dye.

This species is found abundantly in the woods about Belleville and Kingston, appearing to thrive most freely in the calcareous soil of the limestone districts.   It is rare in the woods to the northward and about the small lakes of the Otonabee.

BUTTONWOOD.—AMERICAN PLANE OR SYCAMORE.—*Platanus occidentalis,* (L.)

Button-wood is the common name given to the Plane-tree, by which latter name it is better known in Europe ; but by Americans it is

---

I am indebted to J. B. Hurlburt, LL.D., for the account of the weight of the different woods as given above, and also for many useful and reliable notices from his pamphlet on the "Products of the Forest and Waters of Upper Canada."

generally called Sycamore, the large lobed leaves resembling the Sycamore or broad-leaved Maple. The common term, Button-wood, is derived from the globular heads containing the seeds. These button-like seed vessels remain attached by long thready stalks to the branchlets during the winter.

This noble tree is widely diffused through the western portion of Canada, especially toward Lake Erie and the central townships of the western peninsula. In the rich and fertile lands between the big lakes, it reaches to a great height and bulk, its average height being 120 feet, and 60 feet to the spread of the limbs ; not uncommonly 60 inches in diameter. The wood is hard to split, laborious to chop, and difficult to burn until it has been seasoned for a year or more. The huge trees are cut down and left till the leaves and brushwood are dry enough to help consume the logs. Settlers that left the eastern and northern woods to locate themselves on the more fertile lands in the west, complained greatly of the difficulty attending the clearance of the forest, where these trees abounded, till they learned the necessity of letting the newly cut trees lie till they were fit for cutting up and burning.

The delay in clearing the land was tedious, but it answered in the end. As a general rule we see the largest growth of trees in the richest soil, even the Pine is no exception, as the finest Pines are to be found on the hardwood lands mixed with the deciduous trees ; of nobler growth than those that grow on the Pine ridges and sandy lands.

The bark of the Button-wood exfoliates and falls off in large plates, which distinguishes it from all other forest trees, excepting the shaggy-barked Hickory which also sheds the bark. The Oriental Plane follows the same rule as the Occidental, both trees are often seen in plantations in the Old Country introduced on account of the luxuriant foliage and singularity of habit.

The flowers of the Button-wood are greenish, in dense heads, on long drooping stalks. Dr. Lindley, writing of the Plane, says : " The members of the Plane tribe are natives of Barbary, the Levant and North America. The white wood is valuable. The bark of the Platanus, or Plane, is remarkable for falling off in hard irregular patches, a circumstance that arises from the rigidity of the tissues on account of which it is incapable of stretching as the wood beneath increases in diameter." The Red Pine, *Pinus resinosa*, also parts with the bark in thin round patches, when growing on poor sandy flats. I do not know if this is a general habit of the species. Possibly Lindley's explanation of the shedding of the bark in the Plane and Hickory, may refer to similar cases in other trees.

### BASSWOOD.—WHITE-WOOD.—*Tilia Americana* (L.,

" And humming bees make drowsy music,
  In the flowery limes."

---

" The groves were God's first temples ere man learned
  To hew the shaft and lay the architrave,
  And spread the roof above them."

Our Basswood is closely allied to the Lime or Linden of Europe. It is one of the tallest of our forest trees, and its presence indicates a rich and generous soil, well fitted for the growth and production of wheat, barley, and other cereals. When growing in open ground it becomes a large umbrageous shade tree, forming a magnificent rounded spreading canopy ; unlike most forest trees, it sends up many strong shoots, which, clustering round the parent tree, form groups of vigorous trunks that in course of years bid fair to rival the main central one.

When in full bloom, the scent of the blossoms fills the air with fragrance, and I might say with music, for they form a great attraction to the bees, which crowd the pendent cream-coloured flowers from which they gather an abundant store of honey, murmuring in low bass notes their satisfaction while they labour. The blossoms of the Basswood are of a rich cream-colour, and hang in pendent cymes from the axils of the large heart-shaped serrated leaves ; the fruit is a round, rough woody-nut, 1-celled and 2-seeded ; attached to the peduncle is a large foliaceous bract, which acts as a sail in distributing the seed when ripe, and by aid of which it is borne by the winds and lodged upon some suitable spot, as if a directing power accompanied the winged wanderer on its way, and laid it down where it would find a space to grow and nourishment for the support of the young plant. The Basswood is a quick growing tree, but not very long-lived. The wood is light, white and porous, and though not held in esteem as one of our first rate timber trees, it is yet much used in cabinet and all lighter works ; the wood is soft, close grained and not liable to warp or split. Hurlburt, whom I follow in this description of the Basswood, says : " It is used by musical instrument makers for pianos, also by curriers, shoemakers, and other mechanics ; it turns cleanly, and is much used in the manufacture of bowls, pails, shovels, &c. The weight of a cubic foot is 26 lbs."

The older settlers used the Basswood much in fencing, but the rails are light and subject to decay, becoming brittle and useless after a few years standing exposed to the action of the elements. Pine, Cedar and Oak are more weighty and long enduring, and of course preferable for such purposes.

The green unseasoned Basswood burns badly, and when quite dry too rapidly to be profitable for fuel ; but if it has its bad qualities, it has far more valuable ones to balance them. The inner bark of the tree abounds in bast cells and the fibrous bark is used for mats, the Indians braid it into flexible ropes for various purposes ; they use the inner bark largely in their simple basket work, and have great faith in the healing qualities of the leaves and bark—in scrofulous swellings and other ulcers. The cattle eat the shoots and tender branches, in the early Spring time, when green fodder is not accessible ; and later the large tender green leaves in the forest and sprouts thrown up at the root of the stump, form a favourite repast. In remote, newly cleared forest lands, before grass and straw can be raised for the support of the oxen and cows, the beasts live chiefly on freshly cut branches of the Maple and Basswood, or " browse " as the settlers term such food for the cattle.

The average height of a full grown American Basswood is from 100 to 120 feet, with an average height from the root to the first limbs of 60 to 70 feet, and a diameter of 25 to 40 inches or more.

The Basswood is found all over the Province of Ontario ; westward, and eastward through Quebec. Its range extends to the Lake of the Woods, and it is found more dwarfed in stature, as far northward as Lake Winnipeg, and Norway House. It also occurs in the Valley of the Assinaboine, as ascertained by the botanical researches of Professor Macoun, and from information I have received from friends resident in Manitoba and the North-west.

WHITE-WOOD.—TULIP TREE.—*Liriodendron Tulipifera*, (L.)

> The Tulip-tree opened in airs of June, her multitude
> Of golden chalices, to harmony of birds,
> And silken winged insects, of the sky.—*Bryant.*

The Tulip-tree is only found west of Toronto—not being found anywhere eastward in the Province. It is a noble tree, attaining to upwards of 100 feet, often 70 feet to the first branches, and 36 inches in diameter or even larger. The wood is easily wrought and used for many purposes in cabinet and house carpentry. In the vicinity of the Great Lakes this beautiful tree is chiefly to be found. There is a fine specimen of the Tulip-tree growing at Waltham Abbey, in England, supposed to be of great age. It forms one of the rarities of that place.

The Tulip-tree is so little known by the generality of readers, that I shall give the botanical description, as it may prove both interesting and instructive.

Sepals, 3 reflexed ; Petals, 6 in two rows forming a bell-shaped corolla ; anthers, linear, opening outwards : pistils, flat and scale-form, long and narrow, imbricated and cohering together in an elongated cone, separating, when dry, from each other and from the long, slender axis in fruit, and falling away whole like a samara or key ; one to two-seeded in the small cavity at the base; buds, flat, sheathed by the successive pairs of broad, flat stipules, joined at their edges. The folded leaves bent down on the petiole, so that their apex points to the bud. The classical name is derived from the Greek words signifying *lily* and *tree.*

A most beautiful tree, sometimes 140 feet high, and 8 or 9 in diameter, in the Western States, (where it is wrongly called Poplar.) Leaves very smooth, with two lateral lobes near the base, and two at the point, which appears cut off by a shallow notch ; the corolla is two inches broad, greenish-yellow, marked with orange.—(Gray's Manual of the Botany of the Northern States.)

Hurlburt says, of the Tulip Tree, "it is abundant in the south western counties of Upper Canada ; weight, 30 lbs. to the cubic foot much used in building and cabinet work.

In Professor Macoun's Catalogue, (1883) the following valuable information is given :—"In rich soil throughout the Western Peninsula of Ontario. A noble tree in the thick forest west of St. Thomas, and a beautiful object when covered with its large Tulip-shaped flowers in the middle of June ; cultivated in Prince Edward County, where it flowers freely."

SUGAR MAPLE—HARD MAPLE—ROCK MAPLE—*Acer saccharinum*, (L.)

"Dark Maple where the wood-thrush sings."—*Bryant.*

While we regard the Pine as one of the greatest sources of wealth to Canada, we must not lose sight of the Sugar Maple, the next in commercial value as respects its uses as timber, as fire-wood, its house-hold worth in the production of sugar, and as an ornament to the country by its noble form and rich masses of verdure.

The Maple becomes a beautiful object under cultivation. No longer drawn up to an unnatural height as in the dense shade of the forest, where its outline can scarcely be distinguished from the surrounding trees, it developes into a grand and sightly object, forming a finely rounded head, the long rather slender branches curving upwards and outwards, clothed with rich masses of dark green foliage ; leaves broad, smooth on the surface, divided into three principal pointed lobes and two inferior ones at the base. A great boon to the cattle, that seek the deep cool shade from the noon-day heat of a Canadian Summer sun, is a group of these noble trees : but it is a blessing which the too greedy farmer often denies to his beasts, grudging the space, that is occupied by the trees, from the money producing cereals.

In its native woods the Sugar Maple attains to a lofty height, often in rich soil measuring from 100 to 120 feet, with a circumference of 12 feet. Like the Elm, the head is lost in the general strife for room, and is small in proportion to its great height, wanting the fulness and roundness of outline, which it readily acquires when favoured with space and free access to air and sunlight in the open clearing.

The bark of the Maple is light-grey, smooth, till it attains to the age of from fifteen to twenty years, when it begins to form rifts at the lower portion of the trunk. Every year as the wood increases the bark becomes more rugged, and in the old forest tree it is thick, bluish-grey, and deeply furrowed. Where the soil is wettish, unseemly knots and huge excrescences may be seen on the trunk of some of the old trees. The disease may be caused by injury during its early stages of growth, an interruption of the sap or a puncture in the tissues of the inner bark by insects, but whatever be the cause, while the symmetry of the trunk is deformed, the thrifty backwoodsman, always good at expedients, turns the ugly excrescences to good account and converts these woody lumps into useful beetles for splitting rails, driving posts, and wedges, and similar purposes. A good Hard Maple knot is no despicable instrument in the powerful hand of a Canadian settler in the bush.

The beautiful markings in the grain of the wood of the Maple, forming what is called Birds-eye and Curled Maple, cause it to be highly valued for ornamental cabinet work.

The wood of the Maple is hard, finely grained, and takes a good polish ; it is largely used by the carpenter, the cabinet maker and machinist. The timber is used by the wheel-wright and waggon-maker, when well seasoned, being prized for its durability and great hardness for axles. For firewood the wood of the Maple is second only to that of the Hickory, but being more abundantly distributed throughout the Dominion, it is more generally used and is considered the very best of fuel ; it burns readily giving out a great degree of heat. The ashes yield the best lye for the home manufacture of soap, and for the production of pot and pearl ashes as an article of commerce. The Maple requires a generous soil, and indicates to the settler good paying Wheat growing land ; where the soil is cold, wet and mossy, the trees when they do appear are stunted, thin and scraggy, the sap is weak and watery, yielding a poor return to the sugar maker.

Like the Pine, the Maple has many enemies among the insect borers ; to obtain the larvae, the tree thus infested becomes the resort of the Woodpeckers ; several species of these hardy birds winter with us, making their abodes in the thick forests.

It has been supposed by some persons that it is to obtain the sugary juices from the trees that the Woodpecker bores the bark, but this is a fact not fully established, and may be considered as doubtful, as the bird also carries on its labours at a season when the sap is in a dormant state, and consequently would not repay the trouble of boring. We must give our sagacious birds credit for better insight into their business.

I have sometimes noticed one of the small Downy or Midland Woodpeckers tapping and rapping away at an old dry post, certainly where no sap could flow out to repay it for its work. The bird would stop and seem to listen for a while, and then renew his work. I think he heard the prey within.*

I once noticed a small black and white, red-crowned Woodpecker diligently hammering away at a dry stick of wood lying in the yard, but his exertions were useless. Some time afterwards the wood was split, and a large grub was found in the middle of the stick.

Bees often frequent the sap troughs and regale themselves with the sweet fluid. I had noticed bees coming and going on sunny days in the sugar bush, and on mentioning the circumstance to an old Yankee settler's wife, she told me that the wild bees frequented the troughs for the sweet sap: "I guess them creeturs like good things as well as us humans," she sagely remarked; "I kinder like to see them helping theirselves, and I say to myself, yer welcome to what you take." She was an odd looking old woman, but I was pleased with her benevolent hospitality to the "wild creeturs" as she called the bees.

The northern side of the Maple is generally clothed with a thick coating of moss and liver-wort for many feet upward; probably a provision of the wise Creator for defending the inner tissues of the tree from injury during the season of intense cold, from without, or to prevent the escape of heat, from within. On the southern side the sap flows more readily and earlier than on the northern; but as the sun gains more power and the days become warmer, the settler taps the tree on the northern side and obtains a good flow of sap not inferior to that which was obtained from the first incisions on the sunny side of the tree.

Toward the latter end of the sugar-making season, which generally terminates in the beginning of April, there is a tendency to acidity in the sap, and the syrup will not "grain" well. The latter boilings are made use of for vinegar, or syrup for immediate use. A good home-made wine can be made with very little trouble from Maple sap, and also a cheap and palatable beer with the addition of a small quantity of hops and yeast to "set the liquor to work."

* The only Woodpecker, against which the accusation of being a sap-sucker can, with any reason be brought, is the yellow-bellied Woodpecker, and it would appear that this really does sometimes "tap" trees for the sugary juices.

It is not necessary in this description of the Maple to enter into the process of sugar making, as many writers have described it.*

It might naturally be supposed that by draining off so large a portion of the juices of the tree, that the vital forces necessary for the production of fruit and foliage would be greatly impaired, but the deficiency is by no means perceptible : it would seem as though the tree were endowed with additional strength to meet the emergency, and repair the waste of the life supporting sap. It does not appear that the leaf falls earlier from the tapped Maple trees, than from others that have not been subjected to the same exhausting process, and no perceptible failure of vigour can be observed ; still in the course of years the energies of the tree may suffer, and its life be shortened, though many an ancient Maple stands in the sugar-bush hoary with age, still putting forth its coronal of leaves and its slender dooping pale flowers and winged fruit, though bearing on its rifted bark, the unseemly scars of yearly woundings from the settlers axe or auger.

The Maple has a wide geographical range on the continent of North America. It is found Southward in Virginia, Westward in Wisconsin, and Eastward in the New England States ; while Northward of the Great Lakes, even as far as Lake Superior, it forms one of the grand features of our far-stretching forest-lands. While we consider its many valuable qualities, we can scarcely wonder that the Maple has been chosen as the emblem for the Dominion of Canada.

The Maple tree and the Maple leaf, are seen in all our national ornamental designs—and summing up all this tree's varied claims it seems worthy of the honour bestowed upon it.

In every country some favourite tree, shrub or flower has been selected as a national emblem or as the heraldic crest. We may mention a few from among many others. Thus we have the Rose, Thistle and Shamrock, illustrative of the Union of England, Scotland and Ireland ; long may the national garland remain entwined in our Royal escutcheon.

The Oak is ever spoken of as the English tree, *par excellence.* The *Pinus sylvestris* called Scotch Fir, of Scotland, the Pine of Norway, the Olive of Spain, the Lime or Linden of Germany, the Laurel and the Bay of Italy, the Palm of the East, the Lily of France, &c. While many a Coat of Arms among the aristocracy of England, bears for its crest or quartering, such simple emblems as a sprig of Heather, or Holly, or Yew, of Olive, or Laurel, or Bay. Borne on the helmet, or in the cap, they were the symbols by which the wearer, and possibly all his clan or

---

NOTE—An excellent practical account of the process, from beginning to end, may be found in Major Strickland's book "Twenty-seven years in Canada," also in the "Female Emigrants Guide' by Mrs. C. P. Traill.

vassals, were recognized, and very interesting it would be could we know the particular history attached to these old heraldic bearings, now lost in the darkness of the far away past.

We might picture to ourselves the parting of some brave warrior, bound for the wars of Palestine, with the dear object of his affections reverently placing in his cap, or helmet, a leaf plucked by her own hand from the trysting Oak, or the sprig of Laurel, which was to ensure victory to the wearer. The Cross-leaved Heath, gathered from the Moorland, was a sacred Christian emblem, as also was the Shamrock, emblematical of the Holy Trinity. The Olive leaf betokened peace to the wearer, and restoration. The Holly tree hardihood and endurance. While the armed Thistle spoke of resistence and defiance.

In almost every country some particular species of tree seems to hold a sort of pre-eminence and to flourish beyond its fellows. We read that in Denmark three successive races of forest-trees have held the soil,—the Pine, the Oak and the Beech, in succession. The natural conclusion is that the Pines have exhausted some element in the soil that was necessary for them ; they retire, as it were, from the land, and the Oaks take their place and flourish for a long succession of centuries. The ground again ceases to yield something necessary to support the Oak, or maybe some change in the climate proves unsuited to its growth ; by degrees the Oak groves give place to the Beech, and in some distant future, the Beech in turn will prepare the way for the Maple, or some other deciduous tree. I have, however, already alluded to this. Let us return to our subject, the generous Maple. We have dwelt upon its value to the mechanic, the timber merchant, and the house-wife, for the luxuries it yields in the form of Sugar, Molasses and Vinegar. &c.; also of of beauty of its form and colouring to gladden the eye of taste. Have we exhausted every subject ? No. Yet there is its value as firewood ; and is this all that we can say ? Not quite. Cast your eyes upon the ground strewed in October with that thick carpeting of fallen leaves, so lovely to look upon, that the eye wanders over the beauty of their gorgeous tints, with a feeling of sadness and regret, that in a few short hours they will vanish from our sight. And is this all ? No, those heaps of dying leaves, have a lesson that speaks silently to the thoughtful mind of the decay of all that is bright and beautiful on earth, and its fleeting possessions ; and yet we may also see how the fall of the leaf and the destruction of the worn-out fallen trunks and branches are gradually preparing a rich and fertile soil for man's use and future maintenance. Therefore, in life and in death, we have cause to value the Canadian Sugar Maple, and with grateful hearts to thank the Giver.

There is another variety of this Maple known as the *Acer nigrum* of botanists, or Black Maple, this tree also yields a sweet sap, and is used occasionally, but the Sugar Maple proper, takes precedence of all the species for the purpose of sugar-making.

RED MAPLE—SWAMP MAPLE—*Acer rubrum,* (L.)

" When April winds grew soft,
  The Maples burst into a flush of scarlet flower."—*Bryant.*

The poet's description belongs only to the Red-flowered or Swamp Maple. The blossoms of the Sugar Maple are of a delicate tinge of yellowish green, in long pendent racemes, very graceful indeed, but less brilliant in colour than the early flowering Swamp Maple. Early in Spring, while yet leafless, the branches are adorned with an abundance of bright red flowers on short foot stalks, clustered together. We see these red-flowered trees by the sidewalks of many of our Canadian towns, where their bright colours give a cheerful appearance to the streets, before any green leaves have been put forth on the deciduous trees. The flowers of the Red Maple are followed by the ornamental winged fruit which is tinged with pink, deepening to bright crimson as the season advances. In all stages there is beauty to be perceived. The large masses of foliage afford a refreshing shade during the hot days of summer after the bright blossoms have fallen.

The native and favourite haunt of the Red Maple is on the borders of streams and lakes, often I have seen the surface of the waters blushing red, when the ruffling breezes have scattered the red flowers in a rosy shower upon the lake, and later in the summer, when the pendent fruit hangs gracefully below the bright green leaves, it is beautiful to watch the reflection colouring the still waters. What painter's colours can match the rich crimsons and scarlets of the foliage ; any close imitation of the gorgeous colouring of those dying leaves, would be regarded as a gross exaggeration of nature. Yet there we behold them, year after year, silently falling and strewing the earth with a carpet so richly tinted that it seems almost a sin to tread such beauty beneath our feet, and mar its loveliness.

Not less brilliant are the autumnal tints of the Sugar Maple ; some few leaves will turn to a vivid scarlet at the end of a branch, while all the rest are fresh and green, the effect is as if the tree were putting forth a rare show of bright-hued flowers ; but it is only the fore-runner of rapid decay.

The Red-Maple sap is deficient in sugary principle, being more watery ; and is, therefore, made no use of by the backwoods settlers, as the result would not pay for the labour required to obtain the sugar.

Botanical description of Red or Swamp Maple: " Leaves palmate 3-5 lobed, mostly cordate at the base, incisely toothed, whitish and nearly smooth beneath ; sinuses acute : flowers red crimson, rarely yellowish—on short pedicels—about five together, pedicels in fruit elongated and pendulous ; fruit red, wings one inch long; a handsome tree common on low grounds and swamps, making a splendid appearance in April, before the leaves appear ; often planted as an ornamental shade tree on account of its red flowers.—*Class Book of Botany.*

Soft Maple—Silver Maple—*Acer dasycarpum,* (Ehrh.)

" No tree in all the grove but has its charm,
Though each its hue peculiar."

This species has its habitat most commonly by the banks of lakes and streams, where it forms deep belts along the low lying swampy shores, and sends up its slender upright branches, to a height of seventy or eighty feet ; the ends of the twigs are red, the leaves are darker in colour than the Sugar Maple or the Red Maple ; lobes five, sharply cut and pointed at the ends, and very white underneath, turning in early Autumn to deep crimson, beautifully blotched, and varied with shades of yellow, rose, and green. When agitated by the wind the leaves "turn up their silver linings to the sun," as the poet beautifully expresses it : the fruit is tinged with rosy-red with large diverging wings. There is a strong black dye extracted from the inner bark which makes good ink, with a small proportion of copperas and a little sugar to give it consistency. The sap is never utilized in making sugar. The wood is white and soft, and of rapid growth, and less valuable than that of the other Maples. But from the rapidity with which it grows and its graceful appearance it is the best species to plant in our towns and cities as a shade tree.

There are several varieties of Shrubby Maples which never attain to the dignity of trees. The Mountain Maple, *Acer spicatum,* (Lam.), is common in open thickets, the young leaves are reddish, and the fruit light green, the surface of the leaf rough and veined, three lobed, downy underneath ; the flowers are borne in upright terminal racemes, and appear in June, after the leaves have turned green.

Another of our Shrubby Maples with racemes of showy yellowish flowers, large three-lobed foliage and green-winged fruit, is chiefly found in woods with the preceding, this is *Acer Pennsylvanicum,* (L.) or Striped Maple.

It is not so commonly found in our woods as the other species I have mentioned. The bark is smooth, and striped with light and dark lines, which give the tree a very peculiar appearance, and from which the incorrect name of Striped Dog-wood has been taken. This tree has a very attractive appearance in Autumn, when the leaves turn to a delicate creamy yellow.

### ASH-LEAVED MAPLE.—*Negundo aceroides*, (Mœnch.)

This tree,. which is now grown in many parts of Canada as a shade tree, seems in the North-Western Province of the Dominion to take the place of the Sugar Maple, and the sap is utilized in a similar way for the making of Sugar, as my son, Mr. W. E. Traill, in his letters from Fort Ellice and Fort Qu'Appelle, writes of Maple Sugar made from the native trees—though he adds : " The tree is unlike in foliage and grandeur of size to our Canadian Sugar Maple. The leaves are more like those of the Ash, and from that cause it is known as the 'Ash-leaved Maple.' It is a handsome tree, and is much valued ; it grows on the Saskatchewan River, and Southward, and should the country become civilized at some future date, it will be valuable for ornamental as well as useful purposes, as it is easy of cultivation."

These remarks were made many years ago, when the prospect of the Great North-Western country becoming a field for agricultural speculation was very remote in the minds of the Hudson's Bay Company gentlemen fur traders.

There is a fine grove of these trees on the banks of the river Humber, near Toronto.

### SPICE WOOD—*Sassafras officinale*, (Nees.)

> " Dark Maples where the Wood-thrush sings,
> And bowers of fragrant Sassafras. "

The Sassafras so well known for its purifying, agreeably medicinal qualities and its sweet scented aromatic bark and wood, is a native tree of the Western parts of North America, and the warmer portion of Canada that lies between the shores of Lakes Erie and Huron. I do not think it grows indigenously Eastward or North of Toronto.

The wood and bark of this tree are highly esteemed by the Indians and the old settlers—the former will take long journeys to obtain a supply of the chips and twigs to smoke and chew, the women make tea of the dried twigs, and the Sassafras is one of the simple luxuries of the Squaws.

A friendly Indian woman once brought me a small bundle of the little branches, as an especial token of her regard. The fragrant offering had been brought all the way from the Credit River in the West, and was considered by my old Indian friend as a great gift. She invited me to visit the lodge on one of the islands in the Katchawanook Lake, where they were encamped—saying—"Come see me—drink Sassafras tea—Paddle you myself." But a report of deer, some miles further up the lake caused a sudden movement among the Indians, and the camp was broken up for the time, and when I did visit the island the hospitable Squaw looked much concerned and said " Husband smoke Sassafras got-a-none." In other words it was all gone—so my treat ended in smoke.

A diet drink called " Sassafras beer," is considered very cooling and purifying to the blood in warm weather, but is composed of several native herbs, roots, and sprigs of trees in addition to the aromatic wood of the Sassafras tree.

The Sassafras is found in the western peninsula of Upper Canada. The timber is small and only used in some ornamental work, for its sweet, pleasant scent. The full grown tree, under the most favourable circumstances, rarely exceeds sixty feet in height, and fifteen inches in diameter. It is from the root bark that the fragrant essential oil is obtained.

The flowers which are in clustered racemes, on short reddish stalks, are greenish-yellow, appearing at the same time as the leaves from scaly buds ; the fruit is a hard, blue berry, supported on a fleshy, reddish, club-shaped pedicel. The leaves are often lobed, but generally ovate and entire : twigs, yellowish green, spicy, mucilaginous, healing, wholesome and purifying.

The Sassafras belongs to the same natural order as the Laurel, but appears not to possess the dangerous narcotic principle of the Laurel proper.

There is a pretty spicy shrub, nearly allied to the Sassafras, known as Wild Allspice, Benzoin and Fever-bush, much used by the Natives with their tobacco.

SPECKLED ALDER—*Alnus incana*, (Willd).

" The Alders dark that fringe the pool."

The Canadian Alder is chiefly found on low ground, on the shores of low-lying lakes and the banks of creeks. It forms dense thickets in such situations, scarcely reaching to the dignity of a tree. The strong, thick, knotted roots send up many woody stems, and these thick knots are not without their value, being varied and ornamental in the grain, and are used for inlaying. The hard knobby roots are also used by the farmers for beetles in their field work, splitting wood and fencing. The leaves are also used in dyeing black. The catkins of the Alder are formed in Summer ; the leaf buds may be seen as soon as the old leaves fall. The fertile catkins when fully grown have a pretty cone-like form.

Early in the Spring, often in the beginning of April, before any other tree or plant has shown a sign of awakening from the long winter's sleep, gradually the pollen-bearing catkins begin to lengthen, until graceful yellow tassels, of two or three inches in length, hang from every slender twig. On account of its early appearance this glad harbinger of Spring is a welcome sight to all. Closely following in its wake come the flowers of the Silver Maple, and then the lovely Trailing Arbutus follows with many others of our sweet Spring flowers. The leaves of the Alder are broadly ovate, rounded at the base, coarsely toothed and downy underneath.

### POPLARS.

The Poplars and Willows are among our most common native trees and shrubs. We have a great variety of species of both, from the lofty Cottonwood of the Western peninsula on the shores of Lakes Michigan and Huron, to the dwarf Willows that form the chief portion of the vegetation of the far Northern boundaries of Hudson's Bay.

Little valued as timber in the thickly wooded lands of Eastern and Western Canada, where hardwood trees abounds, such as the Walnut, Maple, Beech, and others, yet in the more distant divisions of the Dominion, where timber trees are less abundant and in the prairie bottoms where the Poplar and Birch form the only trees, the Poplar rises in value and is used not only for firewood, but for building purposes where better timber is not easily obtained, unless at a very great cost. In remote places far from the means of transport, and where, for the present, saw-mills are not in existence, the Poplar supplies the place of a more durable wood. The Romans called the Poplar "the Tree of the People," and it was used to decorate the public walks.

THE COTTONWOOD—NECKLACE POPLAR—*Populus monilifera,* (Ait.)

Is the finest of all the Poplar family. It is a noble-growing forest tree, confined in Canada, chiefly to the more Westerly portion of the Dominion. Between the Great Lakes, seems to be its particular locality. It loves the margin of lakes and rivers, and moist ground, indicating a rich vegetable soil, well repaying the cost of clearing the land of the trees, which are of great height and girth; though the wood is very light when dried, yet it is difficult to burn on account of the abundance of sap while green. On this account the settlers do not attempt to fire the felled trees until a year's seasoning has rendered them easier to burn— so I have been told by an intelligent old Scotch settler who moved Westward some years ago into that part of Canada where the Pine is scarcely seen, and the gigantic Cottonwood and other hardwood trees abound. The wood of the Poplar is used for turning, and carving, and is made into shavings for thin boxes for millinery goods, and for druggists' purposes, and any work that requires very light, easily worked wood.

The foliage of the Cottonwood is smooth, heart-shaped, with prominent nerves, serrate, slightly hairy teeth, the fertile catkins are long, with fringed scales ; the seeds are clothed with white cottony down, from which the familiar name given to the tree is taken. This is the largest and most important of the Poplars. The specific name, *monilifera*, or Necklace-bearing, is applied on account of the appearance of the fertile catkins which are very long, and have the large fruit-pods scattered irregularly along them, giving somewhat the appearance of a necklace. Very common Westward and Northward is the

TACAMAHAC—BALSAM POPLAR—*Populus balsamifera*, (L.)

In favourable situations, in low ground near the shores of lakes and rivers, on gravelly banks and low bottoms, the Tacamahac is found all over the Eastern and North-easterly portions of Canada, filling the air with its balsamic fragrance. It is not one of the largest of our native trees, but where growing in free space it forms a fine bushy leafy head. The foliage in shape resembles that of the Pear tree—large, smooth, ovate, pointed and serrate, whitish and abundantly net-veined beneath.

It is not advisable to plant the odorous Balsam Poplar in the vicinity of gardens on account of its tendency to throw out suckers from the roots, or increase by seeds, which thus become untidy and troublesome to clear away. The catkins of the Balsam Poplar are from two to four inches long, curving, tail-like, and covered abundantly with the white silky down attached to the seeds.

After showers, the leaves emit a fine aromatic odour. The great peculiarity of the Tacamahac is seen in the resinous leaf buds which are formed early in the Fall and are covered with a fragrant yellow gum which is of a most healing nature. It is used as a styptic for fresh cut wounds by the natives, also as an ingredient in a healing, stimulating ointment for bruises and sores. The crushed leaves are also applied to strains and bruises. The buds gathered in Spring and steeped in spirits form a liniment in quality closely resembling the old well-known "Friar's Balsam," and it is equally excellent as an application for fresh wounds being styptic and very healing.

### AMERICAN ASPEN—*Populus tremuloides*, (Michx.)

" Which whisper with the winds none else can see,
And bow to Angels as they wing by them."—*Shelley.*

" Shuddering even without a wind
Aspens their paler foliage move,
As if some spirit of the air
Breathed a low sigh in passing there."

The Aspen may be seen in all low wet flats, forming thick groves of slender growth with greyish smooth bark, which whitens in more open situations. The wood is of little value: it is white, watery, and brittle.

On entering one of these Poplar flats, or *swales* as the country people call them, a sensible change in the air is perceived, the dew seems more heavily condensed, and a chilliness is felt even in warm summer days, while the slightest breath of wind sets every leaf in motion, fanning the air to coolness.

The Aspen is a short-lived tree, subject to a black canker caused by an insect that destroys the bark, and gives an unsightly aspect to the larger trees which are usually found growing on waste lands by road sides, where they spring up spontaneously.

### LARGE-TOOTHED ASPEN—*Populus grandidentata*, (Michx.)

The leaves of this species are large and coarsely toothed at the margins, especially when young; very pointed and downy underneath of a greyish tinge of green above. Like the common Aspen, it springs up on old neglected clearings and waste places. The long, drooping, silky catkins appear before the leaves expand. The growth of the tree is rapid, but becomes unsightly in age with black rifts and scars; the juices of the tree attract the small black ants, and the Wood-peckers help the work of decay either for the juices contained in the tree, or for the insects that take refuge in the bark. The wood is considered of little worth. The buds of the Poplars begin to form early in the Autumn and slowly advance, till in March and April they may be seen swelling the gummy varnished cases which protect the immature leaves and catkins, the bluish silken down peeping out as if to try the temperature of the early Spring before unfolding. Among the low dwarf Willows, the leaf buds often appear so green that at Christmas one might expect to see the bushes clothed with verdure. One is apt to think these premature efforts of the trees are put forth, while the sap

o

is yet active, preparatory to the rest that takes place during the winter months, and yet it is a question if indeed the vegetable life is not always active, though the effects are not apparent.

## WILLOWS.

‘ The Willows waked from winter's death,
  Give out a fragrance like thy breath—
  The summer has begun.’’

Professor Lindley unites the Poplars and Willows in one Natural Order : the alliance seems a very natural one. In many particulars their properties and general habits are alike.

Besides our native Willows, we have others that have been introduced and naturalized in this country. The elegant Babylonian or Weeping Willow, may be seen in many of our gardens and by city sidewalks, but being an introduced tree is apt to be injured by severe frosts. when grown inland and to the north and east, and does not succeed in cold, exposed situations, though it strikes readily from slips. The Willows that are most common under cultivation are the Golden Barked Willow, and the Silver or White Willow. The former is a tree of rapid growth and very ornamental, especially to the Winter landscape, to which its bright golden barked twigs give a liveliness and colouring, when nothing but the white snow and dark sombre evergreens meet the eye.

Like the Lombardy Poplar, the Golden Willow marks the habitation of man : it it a familiar and domestic tree, for it is never seen apart from the homestead. Its rapid growth gives it an early place in the settler's garden or clearing about the dwelling, or in the village street. Many magnificent specimens of this ornamental tree may be seen in some of our older towns overshadowing the sidewalks, and making a grateful shade during the hot hours of the Summer day—cooling the heated air by the incessant play of the slender drooping branchlets and silvery leaves.

The Willow has been used as a symbol of grief from its downward drooping habit. It was worn in token of sorrow by disappointed forsaken lovers, as well as by those who mourned for the death of their beloved. Herrick, our old English poet, says, addressing the Willow,

“ Thou art to all lost love the best,
    The only true plant found !
  Wherewith young men and maids, distressed
    And left of love are crowned.”

While other trees are yet but saplings, the White and Yellow Willows will have attained their full height. The deeply rifted bark of the trunk and large widely spreading branches, bespeaking premature old age. I remember a farmer pointing out to me two grand hoary looking trees—Golden-barked Willows—that overshadowed his house. He was yet in the prime of middle age, without a gray hair. "Those trees," he said, "were planted mere slips by me, when I was a boy." To an immigrant fresh from one of the crowded, smoking cities of the old country, where trees are rarely seen in any of the public streets, the refreshing verdure of the Maples, Locusts, Elms and Willows that adorn the thoroughfares of our Canadian towns must be a source of enjoyment, and prove exceedingly attractive objects. We owe this taste for shading our streets, to the Americans, and thank them for the example they set us, which we now find followed even in our small Canadian rural villages. With the exception of such purposes as hat boxes and baskets, for which the wood of the larger species is used. The wood of the Willow is not very valuable.

A large number of trees and shrubs of this family are indigenous to our country. The two species *Salix alba* and *Salix vitellina*, are said to be of foreign extraction, they are the largest and most commonly grown, but there are several very beautiful species that may be seen on our river banks—*Salix lucida*—Shining Willow is a tall, elegant-growing slender species, with bright-barked sprays and long, very smooth, foliage ; there are too, many lovely dwarf Willows, some not exceeding a few feet in height, remarkable for the light yellow-green foliage and elegant drooping catkins. Some of these pretty dwarf Willows are found on rocky gravelly river-banks drooping over the water. It seems to me that a division should be made between the bright-leaved, bright-barked Willows, and the grey, coarse, hoary upright-growing species, which are neither useful nor ornamental, of this class we have several kinds, as the tall rough-leaved, Grey Willow, with broad, coarse, veiny leaves, upright branches and very long green catkins with dark scales and white down, a tallish tree of little beauty and no value for timber or fuel. Some of these grey-leaved Willows are bushy shrubs from eight to fifteen feet high, branching out from the root-stock and forming thickets on low wet ground, while others are found on dry hills and open grassy plains. There are dwarf species, such as the dwarf Low Bush Willow, *Salix humilis*, which abounds on those open tracts of ground, known as Oak Openings and Plains. These are rather pretty little shrubs, even after the fall of the leaf ; the green leaf-buds and silvery catkins give a look of life and promise of better days to come, of Spring and sunshine in store for us. There is a peculiarty in this small shrubby Willow which is the oval leafy cone-like gall which terminates the branches, it has a pretty effect and remains persistent all through the Summer, and till the Spring, when it hardens, turns grey and falls to pieces. These cones are attributed by naturalists to the

puncture of a gall-fly, (*Cecidomyia salicis-strobiloides, O. S.*) as in the
Rose and some other plants.   I have opened many of these Willow
·cones at different stages of growth, but have been unable to detect any
larvae or fly in any one.   The cone seems to be formed by imbricated
leafy scales, growing round a central axis.

There have been attempts made to introduce certain species of
Willow into cultivation as hedges for field enclosures, but hitherto the
experiment seems to have failed, for unless the young branches are kept
·cut, the plant runs up and becomes too weak to form a trunk, stout and
strong enough to resist cattle, and besides the labour requisite to keep
down the yearly sprouts is very great.   All the Willow and Poplar tribe
seem to have great attractions for insects of various kinds.

There are several species of small Willows that are now cultivated
in Osier grounds for the manufacture of baskets for field work, and also
for ornamental and domestic purposes ; these coarse osier baskets are
taking the place of the Indian baskets that are woven by the Squaws,
and which formerly were in great request by the settlers.   These Indian
baskets were made from the inner bark of the Oak, Basswood and Ironwood, and were cheap and useful ; the Indian women selling them for
flour, meat or vegetables in the way of trade, but now they look for
money for their wares, and the workmanship not being as strong and
good as it used to be, they find less sale for their basket work among
the farmers.

A much stronger, and more durable kind of Indian basket has
lately been manufactured by the Indians from the inner bark of the
Oak and Basswood.   The men even help in making these heavier baskets,
which will last much longer and bear more work in the field and barn
than the lighter sort.

————

This brings to a close the description of the more important of our
forest trees.   It is to be hoped that the efforts which are being now
made to encourage the planting and cultivation of these valuable pro-
·ducts of the country, will be successful ; and I hope before long that an
Arbor Day will be an established and recognized institution in every
Province of the Dominion.

It is to be regretted that we have no Botannical garden in Canada,
where the productions of the country could be collected and preserved.

Our forests, year by year, are disappearing.   Our stately native
trees falling before the force of axe and fire ; in a few more years our
noble Pines will be utterly eradicated, and the names of many of our
loveliest wild flowers, and native flowering shrubs, will be their only
memorials to say that they once existed on the face of the earth.

Our beautiful graceful ferns will no doubt disappear with the forest trees that sheltered them at their roots, and will be lost; their graceful forms may still be found in lonely swamps, and on the shores of distant lake and river banks, by the diligent botanist in his rambles in search of wild plants; but they will be regarded as rare indeed by those who see them only by chance, as curiosities, in the herbarium, no longer growing wild and free in their native verdure and beauty.

In all countries, but our own Canada, there are national parks and extensive botanical gardens, and it seems a strange thing that with such vast materials at our command, of soil and vegetation, that our country should be destitute of an institution so necessary and so valuable. Will not some of our legislators see to this, for is it not a fact that our resources as respects the natural productions of the land, have not as yet been fully recognized? We have wealth that has never been utilized in the form of medicine, dye stuffs, and materials for manufacturing paper, linen, cordage, &c., to say nothing of adornments for our home-steads in climbing plants for our verandahs, and lovely flowers for our garden borders, and native fruits which by culture might be converted into household luxuries. Men from other lands carry home treasures for the greenhouse, the uncared-for products of our plains, our forests and swamps.

The pride of many a gentleman's garden in England is the " American border," where are cherished many of our wild shrubs and flowers. The Canadian can scarcely believe that some of those plants that are admired and valued by the florist, grew wild and uncared for on his own land, possibly trodden down and despised as weeds; but it is true in this as in regard to other matters " A prophet is not without honour save in his own country."

Every plant, however simple, has a history attached to it—a use if we would but seek it out.

Before our woods are utterly despoiled, and the plants they nourished and shaded are forgotten, some record should be preserved of their uses and their beauties.

There are teachings to be gathered even from the grass that we tread upon. Did not Our Lord, Himself show us this, when he spake of the Lilies of the field? So little interest has been felt in this branch of the Natural History of Canada hitherto, that scarcely any of our young people, children of the educated Canadians, know even the common local names of the plants seen by them in their daily walks; they are cut off by ignorance from many sources of simple unalloyed pleasure, a stepping stone, as it were, to higher and more intellectual enjoyment.

# Our Native Canadian Ferns.

" Hie to haunts, right seldom seen,
" Lovely, lonesome, cold, cold, and green."

" Where the copsewood is the greenest,
" Where the fountain glistens sheenest,
" Where the Lady-fern grows strongest,
" Where the morning dew lies longest.
" Hie Away ; Hie Away."—*Scott.*

Under the Natural System of botanical classification, the vegetable kingdom is divided into two sub-kingdoms, viz., Phænogamia or Flowering Plants, and Cryptogamia or Flowerless Plants. The latter of these is again divided into three divisions ; the first of these divisions is distinguished by the plant having a regular stem or axis, which grows by the extension of the apex only ; it is to this division that Ferns belong.

A fern consists of the frond or leafy part of the plant on which the sori or fruit dots are situated ; these sori are made up of clusters of sporangia containing the spores—the seeds, in other words, of the fern. The stem which bears the frond is called the stipe, *i.e.*, the naked portion below the pinnæ or leafy part ; the rachis is the continuation of the stem, which extends to the apex or end of the frond.

The lower part, or subterranean stem is known as the caudex, and by later writers as the rhizome ; the latter we shall adopt, as the former term is usually now confined to the upright stem of the Tree Ferns.

The rhizome is not the true root, but that part from which the fibrils or roots proceed ; the rhizome is an extension of the axis, and bears something of the same relation to the roots as the tap-root of a tree or woody-stemmed plant does.

This rhizome or root-stock is covered in most ferns with a black or brown bark of bitter astringent nature, with coarse scales, and soft chaff or hairy wool. This root-stock often extends below the surface of the ground horizontally, and contains the embryo fronds wrapped up in chaffy or scaly buds, as may easily be seen in most, if not all, of the dorsal ferns—those ferns that have the sori on the back of the frond, the Aspidium family, Aspleniums, and others. If the ferns of this family be examined in the Autumn months, the round, knobby, scaly buds will be seen clustering the rhizome. On opening or dividing these, the perfect green frond will be found clothed with soft brown scales, closely rolled up and ready to expand as soon as the warm breath of Spring has warmed the newly uncovered mould of the forest.

Some ferns that extend a running root-stock along the ground, have the protecting bud sharp and pointed. The frosts of Winter cannot penetrate these coverings. Ferns of this sort are described as being circinate, or folded within the bud in a circular form, and where this arrangement is not present, other means have been provided for the preservation of the future growth.

In the Moonwort or Botrychium family, the fronds are wrapped in a thin membranous sheath, like that which is seen in bulbs, such as the Crocus, Narcissus, and Snowdrops. The roots spring from a fleshy crown, striking down deeply into the soil ; they are stout and fleshy, covered with a tough brown skin ; in *B. ternatum* these roots are furrowed with rings. The sheath protects the young frond till the Summer is already advancing, and there is no fear of frost nor chilling winds to injure the tender plant. In the Osmundas, the root-stock is hard and long-lived ; the same plant enduring through many years, rooted firmly to its native soil, and hard to dislodge without the use of a sharp instrument and strong arm to use it.

The writer knows old plants of the Cinnamon Fern, *Osmunda cinnamomea*, and *O. Claytoniana*, which she became acquainted with more than forty years ago, still growing and flourishing at this date. *O. Claytoniana*, with its beautiful broad waving fronds, used to be known formerly by its descriptive specific name of *Osmunda interrupta*, which indeed graphically described its peculiar habit, distinguishing it from all other ferns ; it seems a pity to change a name so suitable for one that conveys no distinct idea of the character of the species.

The roots of those ferns, which are nearer to the surface, are more carefully guarded from injury by a close covering of scales, so that no cold can penetrate within, as in such tender species as the elegant Winged Polypody, *Phegopteris Dryopteris*, the root-stock of which is horizontal, creeping, and furnished with roundish white scales fine and thin, which one might think would hardly be a sufficient protection to the tiny, slender buds that it puts forth ; but there is always compensation in Nature. This delicate fern is generally found among mosses, under the shelter of the trunks of fallen trees, and covered by a warm coat of fresh fallen leaves and decayed wood, so that it needs not the thick coating of chaffy scales so necessary for the preservation of its neighbours, the Aspidiums and others, that are circinate, (rolled up in the bud), and which lie more upon the surface.

In many of the ferns the stipe or stem is channelled, and strong very elastic nerves may be found on either side, leading up to the rachis and branching so as to form the main mid-rib which supports the leaflets or pinnæ of the frond ; all these veins and ramifications of the veins are extensions of these nerves, while the indusia or membraneous coverings of the fruit dots which protect the organs of reproduction, the spores, may possibly answer the purpose of the calyx or corolla of flowering plants. So carefully has been the view of the All-wise mind to preserve from injury the minute, I might say mysterious, organs of fructification in the Fern tribes, that we see in many species an additional covering, provided by the alteration of the margin of the leaves ; the edges of the lobes being rolled backwards over the sporangia as in the Maiden-hairs and Brackens, or in the *Botrychiums* and *Onoclea sensibilis* and *Osmundas* where the pinnules are closely rolled in like a ball. The fertile fronds of these ferns present a very different appearance to the Polypodies, Aspidiums and Aspleniums. In the Ostrich-feather Fern *Onoclea Strutheopteris*, the leaflet may be distinctly detected in old fertile fronds that have borne the battering of the wintry storms, and the rains of early Spring. It is very difficult to detect the appearance of seedling ferns of the first year's growth, so few persons having had the curiosity to take notice of the tiny things, to search for them in their native haunts, or attempt raising the plants from seed. This can easily be done in a Fern case, or even in flower-pots covered with a sheet of glass if supplied wi h suitable mould as from the forest, and not exposed to the sun. This mode of studying out the life histories of these interesting plants is now extensively practiced by Pteridologists, as those who make a scientific study of Ferns are called ; but the experiment would prove interesting to everyone who had time and leisure for attending to the culture of these charming denizens of the shady grove. The culture of ferns has for many years past attracted the attention of Canadian ladies of taste, but it has chiefly been directed towards collecting exotic specimens, rare and costly ; but few appear to be aware that our own woods, and swamps, and rocks, afford many beautiful species not less admirable than those that are sold by the nursery-man at high prices. *

Many of our fair Fern-fanciers have little knowledge of the treasures hidden away in their own neighbouring woods.

* Last Summer, in my wild garden, I had twenty-two different kinds of Ferns, brought from the woods and swampy ground about the neighbourhood, or islands in the back lakes, within a few miles of my home.

The very names are foreign to them ; the cultivation of Ferns has become a necessity, because it is the fashion, and the possessors have great pride in exhibiting the beauties contained in their conservatories, but really many of them know little of the nature of the lovely plants on which they bestow so much money and attention.    They are costly and are beautiful to the eye, that is all.

"What constitutes a fern and how is it to be distinguished from any other plant," is a question that is frequently asked.    One young lady brought me a leaf of one of our earliest forest-plants—*Osmorrhiza brevistylis*, the herb known as Wood Parsley, or by the pretty name of Sweet Cicely.    She thought it was a beautiful fern, and seemed surprised when I told her she had mistaken the pretty bright green divided leaf of an Umbelliferous plant for the frond of a fern.    While another brought a leaf of the homely Yarrow, *Achillea millefolium*, and was mortified because I rejected it.    She had gone through a course of Botany at school, she said, and ought to know a Fern when she saw it.    I thought so too.    The fact was that my friend had learned the names of the principal organs of a flower, and could tell its constituent parts by rote, but that was all that she knew or cared to know, and laughed at me for my love of "Weeds," as she called all wild flowers.    My little boy had a higher appreciation of the beautiful, and indignantly resented the word "Weeds," as applied to the handful of flowers that he had gathered for mamma, looking at the lady with wide-opened eyes, he said "Not weeds, God's beautiful flowers."    But I am digressing and must return to the description of the distinguishing parts of a Fern. The roots and fibrils of many Ferns are clothed with fine brown hairs, as in those of the Adiantums or Maiden-hairs, and of some of the Polypodies.    These may be easily examined by the use of a magnifying glass.    This fine clothing of the roots may be as a defence from the frosts of Winter, or a means of conveying nourishment through these delicate organs of the plant.

By cutting through the rhizome or horizontal root stock of *Onoclea sensibilis*, our Oak-leaved Fern, a succession of stems and fronds yet undeveloped may plainly be seen.    This ugly hard rugged rhizome is a wonderful repository of beauty and order of vegetable organisms, kept as in a safely locked store-house, to be produced in due season.

The stipe or stem of a fern is the supporting column.    The rachis, the portion of the stipe which bears the leafage, branching off on either side, forming the mid-ribs and veinings of the leaves or pinnæ which are often again divided and subdivided into leaflets or pinnules.

The rachis may be considered as the supporting frame work or skeleton of the frond which holds together and supports all its parts.

In some species the stipes are clothed with finer or coarser scales, which have been thought to be dead or effete leaves ; in some the scales extend along the rachis, and even to the mid-ribs of the pinnæ and pinnules. Some of these clothing scales are soft, others more bristly or hairy ; in some ferns they are light fawn brown, in others very dark ; some are silvery white, as in the early growth of *Osmunda Claytoniana,* and in the Lady-fern, *Asplenium filix-fœmina.* Some lose the scales as they advance to their full growth, others retain them.

As regards their stems or stipes, ferns are found to vary very much ; in some ferns they are round, smooth and polished. Some are deeply grooved down the centre. Some are black like ebony, others green of many shades, red, or of a fine warm brown. That beautiful fern known as Gossamer Fern, *Dicksonia pilosiuscula,* has the stem of a bright chestnut colour in the older fronds, or golden brown in the younger. This is one of the most lovely of our wood ferns.

Botanists give great prominence to the veinings of ferns, by the situation of which and the manner in which they fork or branch, and the positions which they occupy in the leafage, many nice questions as to the separation of Species, and sections of Genera may be determined.

By spores we must understand what is equivalent to seeds in flowers for reproducing the parent plant. The fern seeds, for so in common parlance we will call these spores, are very minute, as fine as powder or grains of dust, and so light and volatile that they would fly away on the slightest agitation of the plant. But nothing is left to accident. We find a beautiful provision made by the all-wise Creator to protect these delicate organs ; the seeds are covered in all ferns by a thin veil or spore-case known as the sporangium which shields the tender things, and preserves them from loss ; and in some instances, besides this, a further protection is given by the indusium, a membranous covering, formed either from the reversed margin of the leaflet, as in the Brakes and Maiden-hair Ferns, or a separate scale-like organ as in the Aspidiums Nature displays a great variety of methods to ensure the safety of her children.

> " Nature all her children viewing,
> Gently, kindly cares for all."

That a very limited number of fern seeds germinate and become perfect living plants, I think may be inferred from the countless number that are produced on the fertile frond of a fern, compared with the few seedlings that become living plants. A single fertile frond of one of our stately Osmundas or Aspidiums, one would suppose produced sufficient seeds to stock an acre of ground, and yet we perceive no such increase. Some necessary check must take place to keep these plants within proper limits or they are borne on the wings of the winds to far distant places.

The most prolific of our Fern family appears to be the *Onoclea sensibilis*, which produces vast numbers of seedling plants year after year, and as this fern grows in great profusion in my neighborhood I have been able to pay close attention to it and its mode of growth.

Our native ferns seem to be divided by their peculiar habits into three groups : 1, those that love the deep shade and rich leaf-mould of the forest, with its cool glades and sheltered hollows, as the Aspidiums and Aspleniums ; 2, those that delight in wet, spongy soil on the banks of streams and low-lying swampy lake shores, as the Osmundas and Onocleas ; 3, those that find a congenial soil in rocky ravines and rugged, mountainous ridges. Few ferns can live on the open, sunny plains like the hardy Bracken, *Pteris aquilina*, which bears, uninjured, the full glare of the mid-summer sun.

As a general thing the fern shuns the glare of sunshine, turning brown and scorching under the influence of heat and drying winds ; heat and moisture are essential to the healthy development of this beautiful tribe of plants. Therefore it is that we find the finest specimens in damp ground and in warm sheltered woods, while for beauty and rare grace of form, the rock ferns are peculiarly interesting, presenting on a smaller scale features more attractive even than the larger and more luxuriant ferns of the forest and the lake-shore, grand as they are in their development of graceful form and richness of verdure. It is as if Nature, to compensate these rare plants, the rock-ferns, for the rude soil and rough elements, had shed over her nurslings additional charms.

Among our native rock-growing ferns the most noticeable are the Rock Polypody, *Polypodium vulgare*; Rock Cystopteris, *Cystopteris fragilis*, a most lovely, graceful, drooping fern ; Rock Brake, *Pellæa gracilis*, a very small but very lovely species which chooses the crevices of the limestone rocks on the banks of rapid rivers as its home, in steep ridges, almost inaccessible to the foot and eager hand that covets the tender green fronds with which it veils the perpendicular rocky face of its cliff-like abode.

The Woodsias love the clefts of rocks where the soil is black, and into which their wiry rootlets can penetrate and withstand the force of winds and rain. The Holly Fern and Evergreen Rock Fern, our finest Polystichums, flourish in stony forest lands.

From the far distant Hudson's Bay Co's Posts of Lac la Biche and Lesser Slave Lake, I have received specimens of several ferns : a very pretty Cystopteris ; a very small evergreen fern, a miniature likeness of *Aspidium marginale*, size four inches, with fruit dots very distinct ; root-stock very thickly coated with pale, chaffy scales. The same fern was also given me from Lake Temiscamingue, and also a small beautiful frond of *Phegoptoris polypodioides*, and *Woodsia Ilvensis*, from the same place ; this last resembles the Woodsias that were found on the rocks at Fairy Lake, òn the borders of Stoney Lake, where, however, this fern is rare, and not so large as the fine specimens that were kindly sent to me by J. Watt, Esq., and by the late Mr. Barnston, Hudson's Bay Co., Montreal. And here I must remark that Nature seems, by the force of a secret sympathy, to draw together those that take an interest in the same pursuits and tastes, creating a kindly and generous feeling even to the unseen and unknown individual, making the Naturalist eager and willing to communicate to such, their treasures and knowledge. The Botanist and Naturalist seem to me, of all men, the most liberal and the least given to jealousy or envy ; any new discovery seems to be a source of pleasure to all ; they rejoice with them that do rejoice, and regard the happy finder of a new species of plant as a general and public bene-factor. Such, at any rate, has been the fortunate experience of the writer, to which she gladly bears testimony. Much more might be said on this interesting subject, but having, to the best of my ability, described the peculiarities of ferns as to the general characteristics of the order, I will now introduce to my readers the native species which have come under my own observation.

ROCK POLYPODY.—*Polypodium vulgare*, (L.)

The Rock Polypody is found in considerably extensive beds on some of the rocky islets, and along the shores of Stoney Lake, and similar places in the North-east and North-west townships, bordering upon that interesting chain of inland waters, known familiarly as the Back Lakes. It is also found in many other similar localities throughout Canada, but generally in rocky and undrained places, where swamp and rocks abound.

It is not usually a forest or wood fern. The rhizome is yellowish, thick and creeping, to a certain extent scaly, sending up stiff upright stipes, smooth and green, rootlets strong and wiry. In its outline this fern is narrow-lanceolate, simply pinnate, veins straight, forked ; sori large, round, destitute of the thin membranous covering, which the botanists term the Indusium, such as distinguishes the fruit-dots of the tribe Aspidieæ, and some others of the dorsal ferns.

These fruit dots or sori in the Polypodies are the distinguishing characteristic of the group. In *P. vulgare* or Common Polypody, they are large, round, and prominent, of a lively brown, or almost orange colour, and being abundant though quite separate, they give a rich and handsome bordering to the deep green fronds; the sori ranged like beads along the pinnæ, have a striking and lively appearance. The veins are blackish and prettily waved; the colour of the frond is of a very rich, dark green, smooth and glossy, but bright green on the underside. Some of the fronds bear fruit at an early stage of growth. I have found tiny fronds of three inches in length with fruit dots on the upper pinnules. This species is not very scaly, a few large, loose ragged scales, which soon disappear, may be found on the early developed fronds, and at the junction of the stipes with the root.

The Summer heat curls up the fronds but rain soon refreshes them and revives their verdure. During the heat of the Summer of 1864, large beds of this fern were seen drooping and fading on the rocky soil, but revived again when the showery weather came. The fruit dots are in greatest perfection in the month of September.

As yet our native Polypodies are confined to four distinct species, and some varieties; but there is little doubt but that this number will be increased as the knowledge of the fern tribe becomes more general. Already, even in remote inland villages, I find persons eager to know something *more* of ferns and native plants; and collections are being formed among our young ladies; and even school children bring handfuls of ferns and flowers to ask their names, and be told how to preserve and arrange them.

This is cheering and pleasant, and should be encouraged among our rising population of *all* classes.

Such knowledge is good and innocent, purifying the mind, enlarging it and leading it upwards from grosser thoughts, and lower tendencies.

How seldom do we see a *real* lover of Natural History, especially cultivators of flowers, a drunkard or a profane swearer? Where the mind of man is thus fed and occupied there is little taste for sensual pleasures. Let us then encourage the study of ferns, and the cultivation of flowers, with an interest in all the productions of the soil, as much as possible among our people. These things are good for the happiness of the young, they lay the foundation for enquiry, and finally they are serviceable to the country, as they tend to develop its resources in vegetable productions, minerals, and animals—for one study leads to others—and as yet there is a large field open to discovery, and it is open and free to all who choose to use their eyes and exercise the powers with which God has gifted even the weakest among us.

## MAIDEN-HAIR—FAIRY FERN—*Adiantum pedatum* (L.)

This truly elegant fern is widely diffused over our forest lands. You meet with it in the deep, rich leaf-mould of the Beech and Maple woods ; you see its graceful fronds trembling with the lightest breeze on the banks of inland creeks, or growing in tufts at the side, and near the shelter of, half decaying trunks of trees and mossy roots. It languishes and fades in open sunny exposures, loving cool shades and the sheltering boughs of forest trees better than the glare of sunshine and withering winds, so delicate and so fragile are the young fronds.

When I first saw this lovely fern, I gave it the name of Fairy Fern, never having even seen at that time the British Maiden-hair, its prototype ; and the name, so appropriate, has since become popular, and as we court nationality for our pretty Canadian fern, we are unwilling to confound it with the foreign species, and so continue to call it Fairy Fern, a name so well suited to its graceful form. One could almost fancy that Oberon and his Titania had held their moonlight revels beneath its polished stem and verdant shade, that is if we were disposed —as of course we are not—to believe in the existence of the tiny elves, or " the good people " as I have heard some of our Irish settlers call the fairies. A large full grown frond of *Adiantum pedatum* will some-times measure a foot across and two feet in height, but more commonly they do not attain to more than half that size.

The round polished stipe is forked at its upper part, dividing into two equal branches, these are again sub-divided into long slender shafts decreasing in length, so as to give a semi-circular outline, or rather two-thirds of a circle, to the frond ; each of the slender divisions bears numerous almost horizontal pinnules set upon very short footstalks ; in some fronds these little footstalks are very short, so as hardly to be observed, when the pinnules appear more crowded and almost sessile, in others they are longer and give a more expansive appearance to the frond. The upper edge of the pinnule is cleft or cut into by straight gashes ; these again are toothed ; each little pinnule or leaflet is thus sub-divided into three or four sections, which in the fertile frond are rolled back and form an indusiate border over the crowded sori. At first this border looks white, but later in August it takes a very light yellowish-brown tinge. The older and fertile fronds are larger, stouter, and usually of a fuller green than the barren fronds. The root is black, and fibrous ; the fibres finely clothed with a very delicate soft brown wool, which may be seen even without the aid of a magnifying glass. The frond is circinate in vernation ; all the delicate leaf-stems and leaflets being rolled up when they first shoot up from the root, after a few days' exposure they flutter out, as a newly hatched butterfly shakes its wings,

and very soon expand their delicate leafage to the action of the air. At
the base of the stipe and a little way up the glossy stem are thin, ovate,
light-brown scales, but these soon disappear as the young plant gains
hardihood, and we see no more of them. The very young fronds, those
of the plants of a few years' growth, are exact miniatures of the older
fronds, and lovely fairy-looking things they truly are. I think that few
of our perennial rooted ferns bear fruitful fronds before the third season
at the very soonest, some probably still later, which I think is the case
with the Osmundas and Onocleas. Elegant as is the English Maiden-
hair (*Adiantum Capillus-Veneris*), it is not more beautiful in colour or
graceful outline than our Canadian species.

*Adiantum capillaire* is the American name for our Fairy Fern,
which is given under the supposition that it is the plant from which the
old French settlers extracted the famous Capillaire, used as a pleasant
cooling drink, though now almost out of use.

The classical name is derived from a Greek word, signifying—without moisture. The surface of the frond is never wet, as neither rain nor
dew will lie upon it. The seed of the *A. pedatum* is difficult to obtain ;
it is shed as soon as ripe, being as fine as the finest powder ; if cultivated
I should think it should be, by taking up young plants of the first or
second year's growth, with a sufficiency of the black leaf-mould in which
it delights to grow. The only variety I have found in this fern is one
where the pinnules are shorter and broader, and the colour deep sad
green ; the foot-stalks very short and closely set, so as to give a crowded
over-lapping aspect to the frond ; the outline is more arc-like or hemi-
spherical, and the main stem shorter. I have gathered several specimens
at different times, but the form is not common : the age of the root-
stock, and variety in the soil, will give a difference in colour and growth
—but, generally speaking, there is more constancy in the appearance of
this charming fern than in many others. Professor Lawson, speaking
of the Canadian Adiantum, says, fine as it is in Canada he has seen
specimens from Schooleys Mountain where the semi-circular frond
measured two and a half feet in the radius.

" In the days of the old herbalists," says G. W. Johnson in his
work on British Ferns, " the true Maiden-hair Fern was considered not
only efficacious in many diseases, but especially potent in promoting the
length of ladies' tresses, and to this attributed power it owes its name
both among the Latins and the moderns." He gives the following
recipe for making the celebrated syrup called Capillaire :

" Maiden-hair leaves, 2 ounces ; Liquorice-root, peeled and shred,
2 ounces ; boiling water, 5 pints. Let them stand for six hours, strain

PLATE VIII.

I. SCARLET LOBELIA (*Lobelia cardinalis*).

II. ARROW-HEAD (*Sagittaria variabilis*).

and add thirteen pounds of loaf sugar and one pint of orange flower water."—*Johnson's British Ferns, p.* 11–12.

It appears more probable that the familiar name "Maiden-hair," given by the gallant old herbalists of former times, was derived from the black-shining hair-like stripes, or from the soft brown covering of the young rootlets, than from any imaginary virtue in the plant for promoting the growth of the human hair. That singular and beautiful little plant *Spiranthes gracilis*, owes its pretty name of Ladies' Tresses to the spiral arrangement of its delicate pearly-white flowers, on the twisted stalk, which suggested the idea of the ringlets of hair on a woman's head.

### COMMON BRAKE—BRACKEN—*Pteris aquilina*, (L.)

Though found growing so abundantly on dry, sunny wastes, and, therefore, considered by many persons indicative of a poor, sterile soil, this fern may also be seen flourishing exceedingly, in richly-wooded thickets, and even penetrating within the interior of the forest ; proving the fact that though it will live and grow in light and poor soil, it thrives far better in a more generous one, where its rank, deep green, widely-expanded fronds attain three times the width and height that they do on that which is sterile.

Were it less common it would excite our warmest admiration, from its finely developed branching or fan-like outline, rich colour and abundance of fine coffee-brown sori. There is, too, a great variety, both in colour and shape, of the fronds ; some are of the most delicate tender tint of green, others dark and glossy with purplish stems of various shades, while some are of a rich grass-green, or again, a yellower tint or bronze prevails.

The usual form of a full-grown frond is almost triangular, divided into three spreading bi-pinnate branches, in some the lower pinnules of the pinnæ are twice or thrice deeply toothed, and then terminate in a long blunt, narrow tail-like end ; in others the divisions are crenate or simply lobed ; and in occasional plants the pinnæ and pinnules are crowded on the rachis and mid-rib, crisped and standing forward, bluntly toothed at the edges ; the whole frond wider than long, slightly pubescent beneath, and the stipes either short and thick, or the divisions on long stalks, wide and spreading. This variety seems to correspond in some particulars with the form *decipiens* of Professor Lawson. I have found it in fruit as well as in the barren state. Also another form, simply lanceolate, not branching ; the fruit confined to the lower halves of the pinnules—not extending to the ends of the lobes —probably merely a chance variety, though not very rare, as I have

P

found it in several localities.　Though the Bracken appears to be spread over a great portion of the globe, it is singular how few know of the benefits man can derive from it.　The beasts of the field even leave it untasted; it affords indeed shelter to many of the smaller quadrupeds, the Hare and some others, and the Partridge, Ptarmigan and Quail find a hiding place for their brood among its sheltering stems, in their wild moorland haunts; even the insect tribes seem to leave it untouched. The gardener however, avails himself of the leaves for packing and storing fruit, but still the supply far exceeds the demand.　Formerly its nauseous astringent roots were concocted into an unpalatable dose for children supposed to be troubled with worms, but even that has been abandoned for other more efficacious medicines.　As litter for cattle-yards, it has sometimes been gathered, and the country folk in one of the English counties make a washing lye from the ashes of the stalks and roots, which they use as a substitute for soap or soda.　A bleaching alkali might very probably be obtained from the burnt ashes of this plant that might be found valuable to the manufacturer.

A species of Bracken is used by the New Zealanders as food, as has been asserted by travellers, and the young fronds and root-stocks of our own species can also be used in the same manner.　Though so stout and rigid, the Bracken is one of the first of our ferns to succumb before the influence of early frosts; its verdure departs even before the winds and rains of October have scattered the leaves from the forest trees.

The classical name *Pteris aquilina*, is said to have been derived from the fancied resemblance that the root, when cut transversely, bears to the heraldic figure of the spread eagle—from *pteron*, the Greek word for a wing, and *aquila*, an eagle.

But country maids, when I was a child, read yet more interesting symbols in the cut fern root, and many a time have I seen them poring over the cut portion to decipher the initial letters of their sweethearts' names or their occupation, and wonderful indeed were the auguries they drew.　Here were a swarm of bees, and a hive, that denoted plenty; a plough, that was his calling as a labourer; a bill-hook, or a grove of trees, a hedger and ditcher, or a wood-man; a man casting seed into the ground was a master farmer, and various other devices were conjured up by the learned in such sort of fortune-telling—at all events it was as satisfactory as laying out the cards, or tossing the grounds in a tea-cup, and somewhat more picturesque.

### Rock Brake—*Pellœa gracilis*, (Hook).

Of this genus we have in this part of Canada two species, *P. gracilis* and *P. atropurpurea*, both of which affect a rocky soil, more particularly that of limestone rocks. In the mountains of British Columbia and Quebec the rare *Pellæa densa* is found.

The first of these beautiful ferns, *P. gracilis* may be found on the western bank of the Otonabee, close to the village of Lakefield, North Douro; its short, tufted, wiry, roots, closely wedged within the crevices of the limestone, from which it is difficult to dislodge them, without destroying the delicate foliage of the plant.

The chief nourishment of the plant must be derived from the moist atmosphere in which it grows; watered by the ascending spray and mist, from the fast flowing river below the rocks where it makes its home.

This species of Rock-Brake is the smallest and most fragile of all our native ferns. Graceful in outline and almost semi-transparent in texture, of a light and tender green colour, it may be seen by a close observer, early in the month of June, clothing the otherwise barren surface of the rocky wall that bounds the river in front of Strickland's saw-mill, and enlivening it with its delicate verdant tufts of foliage.

The fertile frond is somewhat duller and sadder in colour, and stands upright above the drooping sterile fronds. The tallest of these upright fruitful fronds rarely exceeds nine inches, oftener from four to six. The yellow, creeping and tufted root-stock throws out a vast number of fronds. The thread-like stipe of the sterile fronds is so lax that it is scarcely able to support the thin leafy frond and thus causes its drooping habit. The fruit-bearing frond is more substantial, the pinnæ, irregularly, bi-pinnatifid, from five to six or seven pairs, the upper pinnule longer than the two lateral ones, blunt, and bearing the sori in a marginal row, terminating the tips of the forked veins, the margin of the pinnules forming a protecting indusium, at first white, then turning to a very light-brown. The seed seems to come to perfection early in July. In August the plant withers with the continued Summer heat, sheds its fine yellow dust-like spores and dies away.

This very lovely little fern is rare in this part of the country. I have only found it in one locality, but it appears, not unfrequently, on the banks of the rapid Moira, above Belleville, at Ottawa and Quebec, and probably in other similar localities. Truly it well deserves its specific name *gracilis* from its drooping fragile nature and slender habit.

### CLIFF BRAKE—*Pellæa atropurpurea*, (Link.)

Like the former this fern grows in rocky soil. The specimens that I saw came from the vicinity of Hamilton, and also from the rocks below Niagara Falls. The largest of these specimens did not exceed nine inches, the smallest four inches.

The stipes of *Pellœa atropurpurea* are of a fine dark purplish colour very smooth and finely polished, a few broad loose scales appear at their junction with the rhizome, and, when very young, along the stipes, but these soon drop away. The fibrils are black and wiry. In colour this fern is of a dull dark green, very pale underneath. The fronds are leathery in texture. Divisions of the frond broadly linear or oblong, blunt at the end ; simply pinnate, but the lower pairs often eared or bluntly lobed. The sori abundant, forming a continuous line ; edge of the pinnæ rolled back so as to form an indusiate cover to the crowded sori. The lower pair of pinnæ are often deflexed so as to give a somewhat heart-shaped form to the frond. The evergreen fronds remain persistent through the Winter, but many of them turn rusty-brown towards the Fall. I should think this plant rather rare ; it is nowhere to be seen in my own neighbourhood, though possibly it may be met with in limestone ridges near rapid water in other localities than Niagara and Hamilton.

### Chain Fern—*Woodwardia Virginica*, (Sm.)

This tall handsome fern is found only in wet Peat bogs where it throws up its large stiff fronds in the month of July.

The fertile and sterile fronds are alike in form, being as a rule from two to three feet high ; pinnate, with numerous lanceolate pinnatifid pinnæ ; the segments long. The fertile fronds bear an abundance of fruit and have a very handsome appearance. The fruit dots are oblong and linear, and are arranged in chain-like rows parallel with the mid-rib, and close to it. At first they are all distinct and separate, but after maturity they all touch and the chain-like appearance is lost. This fern is rare in most parts of Canada. From the nature of its habitat it is not an easy fern to cultivate. The root-stock is large, yellow and creeping beneath the mosses and other bog plants, amidst which it grows.

### Lady Fern—*Asplenium Filix-fœmina*, (Bernh.)

There is something of poetical beauty in the familiar name, Lady Fern, which will always possess attractions for those who associate its drooping, tender loveliness, with shades and streamlets, woodland glens, and dingles in Britain　How naturally do the wild, sweet lines of Scott return to one's memory, when we name the Lady Fern. How, in imagination, do we

> " Hie to haunts, right seldom seen,
> Lovely, lonesome, cold, cold, and green ;
> Where the Lady-fern grows strongest,
> Where the morning dew lies longest.

The Lady Fern of our forest haunts, is a graceful, lovely, fern of a bright tender tint of green and delicate of texture, withering quickly after being plucked ; not bearing to be roughly handled or exposed to any sudden change of temperature, which soon robs it of its freshness and elasticity. In height, the largest fronds vary from one to three feet ; the root-stock is thick, black, and knotty, the stipe blackish at the root-stock, perennial, chaffy, and the frond circinate. When it first issues forth in the month of May, the underside of the tender frond is whitish, from numerous minute silvery hairs, which disappear after the plant is more fully developed.

In the var. *molle* they are seen giving a soft and somewhat silky look to the pinnules. This form is more lax and drooping than the common typical form.

The fruit dots of the Lady Fern are placed slantwise on the back of the veins, nearer to the mid-rib than the margin ; indusia at first white, turning reddish brown in maturity, long, narrow, sac-like, and pointed at each extremity ; slightly curving inwards ; the pinnæ are numerous, long pointed ; pinnules oblong, toothed at the margins ; the margins of the fertile pinnules are slightly concave ; the rachis is pale green, slightly channelled. The habitat of this elegant fern is in damp woods, at the roots of old trees, often associated with *Aspidium spinulosum*, or at the margin of forest creeks among other moisture-loving ferns.

Several varieties of the Lady Fern are met with in the Douro and Smith woods, about Lakefield, and are common in other townships North of Lake Ontario.

The Narrow-leaved Lady Fern—*A. Filix-fæmina*, var. *angustum*, is a distinct form. It differs in many respects from the type of *A. Filix-fæmina*. The colour is a dull deep green ; the stipe is nearly double the length of the leafy frond, the pinnæ are more distant, narrower in outline, much divided ; sori very abundant, so much so indeed as to cover the entire inner surface of the fertile frond, and contract the pinnules considerably. From its narrowed, contracted, appearance, this species has got the name of Skeleton Fern. The rachis and stipe are strongly nerved and channelled ; not very chaffy at the root-stock, but hard and black. This attains to the height of three feet, in wet spongy ground. There is a great tendency to assuming a tasselled form. Sometimes whole plants will present this appearance, the upper pairs of pinnæ being gathered into rosettes, so as to shorten and alter the form of the frond. In a beaver meadow, near Preston's Woods, Lakefield, I found numbers of fronds thus deformed, and it appears to me that it is caused by a diseased condition of the rachis, probably from the puncture of some insect, which contracts the nerves of the mid-veins of the pinnæ.

Another variety of *A. Filix-fœmina* occurred on a waste, wettish piece of ground near Lakefield. The fronds were short and somewhat broad, stipes very short, black at the lower portion, near their junction with the stock ; pinnules much toothed, slender-pointed, very pale green, and the small light brown seed dots covering only the lower half, close to the mid-rib, leaving the upper portion entirely free. The pinnæ are confluent on the rachis. The young, or rather early, fronds are white and silvery on the underside.

On the same ground I found repeatedly a form of var. *angustum*, the pinnules of which were folded together so that only one-half of each was seen ; this occurred both in the fruitful and also in the barren state of the frond. I have also noticed the same thing in plants of *Cystopteris bulbifera*, but less commonly—probably in both cases the alteration may be referred to accidental circumstances.

The texture of the leaf in *A. Filix-fœmina* is very thin and delicate, and the plant seems to be more attractive to the insect tribes than many other ferns—towards the latter end of the Summer it is difficult to meet with perfect fronds free from the mutilation of the leaf-cutters—for I imagine the depredation to arise from some of the small bees that make use of the leaves and flowers for covering their tapestried cells. I am not certain,—I think so, from noticing that occasionally only one, two or three leaflets are neatly cut off, as with a clean, sharp instrument ; whereas, if bitten for food, the ragged and uneven edges would have been left, or the nerves and veins, as most insects reject them. It is possible, however, that this depredation may have been the work of slugs.

There is a very handsome variety of the Lady Fern occasionally met with in our woods, in which the distinguishing feature is the lengthening of the toothing division of the lower pinnules, forming a sharp, rather prominent lobe, nearest to the rachis.

Professor Macoun has a very fine specimen of this variety in his herbarium.

SILVERY SPLEEN-WORT.—*Asplenium thelypteroides*, (Michx )

This species is rather rare in our northerly townships of Smith and Douro, but occurs more freely in Hastings and other easterly places.

I found a vigorous plant of the Silvery Spleen-wort growing in an old marshy meadow at the root of a stump, among a wilderness of Onocleas and Ostrich-feather Ferns and coarse wild grasses, on the shores of the Otonabee River. It is a coarse-growing, robust fern, with stout stipe and root-stock ; pinnules blunt ; veins free, as in other Aspleniums.

The white linear indusia and silvery hairs are the distinguishing features of this species. It appears to love marshy soil, or rich, damp woods where, among shrubs and coarse herbage, it grows to about two feet in height. The colour is deep green, leaflets blunt, fruit dots, of silvery-white lines, long and narrow, giving a striking effect to the under surface of the frond. The fertile fronds are taller than the barren.

EBONY SPLEEN-WORT.—*Asplenium ebeneum*, (Ait.)

The Ebony Spleen-wort is found on the Laurentian rocks, near Shannonville station, where it was found by that diligent botanist Prof. J. Macoun. It is a rare fern, and one of great beauty ; well suited to green-house culture. The root-stock is black, fibrous and matted ; the tufted stipes short, and, with the rachis, blackish, shining and pliant ; the frond tapering in outline both below and at the apex ; the pinnæ being much larger towards the centre of the frond ; pinnæ 20 to 40, blunt, slightly auricled, toothed, or minutely serrated at the margins, lower pairs very short but increasing upwards. Sori placed slantingly nearer to the mid-vein than the margin, on the straight, simply-forked veinlets. Indusia at first white, but when ripe of a light brown, linear in shape ; colour of the frond, of rather a darkish green ; varying in height from four to ten inches ; simply pinnate ; habitat chinks and clefts of Laurentian rocks ; never occurring, that I am aware of, in woods and forests, but probably will be found among the rocky districts northward of Belleville and the rocky ridges in Madoc, Marmora and the more north-easterly parts of Canada, at present a tract of country not much explored by the Pteridologist.

BLACK SPLEEN-WORT.—*Asplenium Trichomanes*, (L.)

This rare little Rock-fern also belongs to the Laurentian formation. The specimen now before me was found in rocks near Shannonville ; this place appears to be rich in the rarer species of Asplenieæ.

In habit and general appearance *Asplenium Trichomanes* resembles the British fern of the same name, and is considered to be the same species. In height it varies from three to nine inches ; possibly fine luxuriant plants may exceed this standard. It is narrow in outline, upright or slightly curving, simply pinnate, pinnæ roundish or irregular, oval, slightly crenate, some of the lower pairs unequal, but hardly so much so as to be lobed, attached to the rachis by a very slight petiole or hair-like foot-stalk, in some cases hardly perceptible. Rachis, purplish-brown or black ; stipes, below the leafage, short ; pinnæ extending almost to the root-stock, which is matted, black, and fibrous, as in *A. ebeneum* ; sori few ; indusia flattish, pale brown, ripening late in July or early in August.

The Rev. David R. McCord, in his notes on Canadian ferns gives the locality of *A. Trichomanes,*—Chatham on rocks in large clumps, observed in no other locality in Lower Canada. Since this record, however, it has been found in many places in Canada extending from the Atlantic to the Pacific.

GREEN SPLEEN-WORT.—*Asplenium viride,* (Huds.)

This delicate, lovely little fern has been found at Owen Sound by Mrs. Roy, and at Gaspé by John Bell, M. D. In general features and habits it seems to differ very little from *Asplenium Trichomanes.* The chief difference seems to be that the rachis is green instead of black, the pinnæ are ovate, deeply crenated, and the sori are more abundant.

NARROW-LEAVED SPLEENWORT.—*Asplenium angustifolium,* (Michx.)

This handsome fern is found in rich woods ; it is simply pinnate, pinnæ two to three inches long, very slightly petioled on the rachis, minutely toothed, narrowly tapering to a point ; texture thin and delicate ; veins straight ; sori numerous but distinct, arranged in slanting lines close to the mid-vein at the base of each vein ; indusia narrow-pointed whitish. Specimens of this beautiful fern were lately sent to me from Ottawa. The pale, thin leaves, almost semi-transparent, are elegantly contrasted with the rich lines of sori slanting from the base and sides of the mid-rib, and are of a bright rufous brown ; the fruit-bearing pinnules are simple, broad at their junction with the rachis, tapering to the point, the fertile frond being much longer than the sterile. The whole frond varies in height from one to two feet ; roots black and fibrous. This delicate fern is of a very light green colour, and from the fineness of its texture early succumbs to frost. Its habitat is shady, damp woods ; in some localities not rare.

THE HARTSTONGUE—CENTIPEDE FERN—*Scolopendrium vulgare,*
(Smith.)

It is but few who have had the pleasure of seeing this handsome fern growing wild in Canada. Until quite lately, only one locality had been discovered in North America—at Owen Sound—where it was found growing in damp crevices of limestone rocks. The evergreen fronds are very different in appearance from most ferns, being elongated, undivided tongue-shaped leaves, from a few inches in length to two feet. The stipe is generally about one-third of the length of the entire frond. The blade of the frond is in large specimens, about two inches across at the widest part, which is the middle, from which place it tapers to a point at the apex, and narrows towards its base. At the base of the leafy portion, where the rachis begins, are two ear-shaped expansions, one on each side. The rachis is very prominent, and from it run towards the margin of the frond several free veins, once or twice forked. It may not be amiss to state that the term, free, as applied to a vein means that after it leaves the mid-rib— however much it may itself become forked—it does not run into any neighbouring veins, but runs free to the margin of the frond. The veins run somewhat closely together and slant upwards a little; on some of them, on the under side of the frond are the lines of fructification. These are very conspicuous, they are generally about an eighth of an inch apart and are elongated clusters of sporangia, covered by indusia, which, when the spores are ripe, split down the centre and curl back, exposing the abundant sori. The elongated clusters of sori vary in length according to the width of the frond; they cover about two-thirds of the space on each side of the rachis and are borne on the upper part of the frond and about mid-way between the rachis and the margin. It is to this peculiar mode of fructification that the genus owes its name. *Scolopendra* is the Latin name of a kind of Centipedes, the many legs of which the dark brown parallel lines of fructification, on each side of the rachis, are supposed to resemble. The stipe is of a dark purplish colour and in large luxuriant specimens, this colour is frequently carried on along the rachis. The Canadian specimens are generally of a lighter green, and the fronds thinner than the English. The plant, too, is smaller. In England, where it is a very common fern, there are a great many varieties, some of which are crested and cut up into many lobes at the end of the frond, or waved and crimped along the edges. This fern takes kindly to cultivation, and grows luxuriantly in a fern case or cool conservatory.

WALKING LEAF—WALKING FERN—*Camptosorus rhizophyllus*, (Link).

In many respects this rare Rock-fern resembles the Hartstongue very closely, in fact it has been described by some botanists as belonging to the same genus. It has evergreen, entire fronds, with auricled or hastate bases; the apices, however, of the fronds are very much attenuated and elongated, and when fully mature curl down and take root at their tips. It is from this characteristic that the popular name of Walking-Fern is taken, because the fronds, rooting at the tips, develop into new plants, which again throw out fronds, which go through the same process, and in this way the plant "walks" farther and farther away from its original source. I have sometimes found plants of three generations all joined together in this way. The fronds grow out in all directions from the crown of the tufted root-stock, and are more or less procumbent or appressed to the rocks on which the plant grows. The fructification consists of oblong or linear sori, irregularly scattered on either side of the netted veins of the fronds, those nearest to the mid-rib single, the outer ones generally in pairs, which run together at their ends and form crooked lines. The classical name is derived from two Greek words, which mean—curved—and—a heap, or fruit dot, in allusion to the manner in which the exterior fruit dots run together. This fern, although rare, is pretty generally distributed over Canada, and is not at all uncommon on the limestone about Ottawa, and at Owen Sound.

BEECH FERN.—HAIRY POLYPODY.—*Phegopteris polypodioides*, (Fee.)

The distinguishing characteristics of this fern are its triangular outline, creeping root-stock, and brown, pointed, chaffy hairs, which may be seen on the rachis and on the veins and veinlets at the back of the frond. Sometimes the lower pairs of pinnæ are much deflexed, but not so decidedly so as in the next species, *P. hexagonoptera*, which is less chaffy or hairy on the rachis, and the two lower pinnæ droop lower than the upper ones. Sometimes, however, it bears them on the left side of the rachis, upright or slantwise as if held up. In a large bed of these triangular shaped ferns, this peculiar arrangement occurs in a remarkable manner. It is only a frequent variation from the more usual form, as it occurs indiscriminately on the same root-stock with fronds with the deflexed pinnæ. The same variation is also, though more rarely, found in the Beech Fern now under consideration.

The pinnules of *P. polypodioides* are lanceolate, blunter at the points, bluntly toothed, confluent on the mid-rib ; pinnæ not so long as in *P. hexagonoptera*, but sharply pointed, colour of the frond a rather dull green, somewhat downy on the surface. The sori small, abundant, but not confluent ; pale brown, when mature, which is in July and August ; situated at the base of the lobes of the pinnules, in lines of twos or threes.

In large-sized fronds, which vary from a foot to eighteen inches in height, the pinnæ form about eighteen or twenty pairs ; the stipe, which is slender and brittle, is nearly double the length of the frond. The root-stock is rather slender, creeping, young fronds very white and downy when they issue from the soil and unroll ; this takes place late in May or early in June. The usual habitat is in rich black vegetable mould in tracts of Maple and Beech woods. I have found this fern near Lakefield, in the Township of Smith, about a quarter of a mile from the river Otonabee, on some waste, half-cleared woodland.

A few years ago this spot was run over by the fire, and only a few starved scorched plants remained, but the following Summer they appeared again in great numbers, but very few fertile fronds in the large bed were to be found. The change of soil caused by the action of the fire and want of shade seemed to affect the texture of the fronds ; they became thicker in substance and leathery, darker and duller in colour, and dwarfed in size.

I have seen specimens of this species differing greatly in outline, size and general appearance, from our Lakefield fern, being much coarser, larger, more lanceolate than triangular ; the lower wings rather curving upwards than depressed, with the sori much larger and more crowded and numerous. Probably difference of soil produced the change in the habit of the fern.

BROAD BEECH FERN—*Phegopteris hexagonoptera*, (Fee).

This species bears a strong resemblance in outline, and some of its more general features, to *P. polypodioides*, but differs from the latter in the lighter, thinner fronds, and the more scanty, hairy scales on the rachis and veins. The sori are round, situated at the base of the lobes of the pinnules. The lower pairs of pinnæ are usually much deflexed, or are both borne on the left side of the rachis at a slanting angle. In this position they have somewhat the appearance of a bird's wings raised for flight. The root-stock is black, creeping and scaly, sending up from pointed buds, at intervals, fronds which are closely rolled up, and covered with a small soft greyish-white down, which after a few days disappears.

This fern forms extensive beds in the soil of rich shady Beech and Maple woods ; height of the stipe, about a foot to eighteen inches. The triangular, wide-spread frond, is about as wide as long ; lobes of the pinnules bluntish ; pinnæ sweeping, curved and pointed at the ends, confluent near the apex, both at the summit of the frond and of the pinnæ.

This is a very handsome fern, but soon yields to the withering effects of early frosts. It dries well, keeping its colour tolerably bright, if not pressed at too early a stage of growth. The root-stock is perennial as in others of the genus. By far the most attractive fern of the group, is that charming fern,

WINGED POLYPODY—OAK FERN—*Phegopteris Dryopteris*, (Fee.)

This is also termed Triangular, and Ternate Polypody The rachis is divided into three parts or branches, which are again pinnately divided ; pinnæ, lanceolate, spreading ; pinnules, crenate or bluntly lobed ; stipe and rachis dark coloured, smooth, brittle, with a few loose thin pale-coloured almost transparent scales, which are not very apparent as the plant increases in size ; fruit dots at the base of the lobes of the pinnules.

The texture of this elegant fern is thin, and it is of lovely shades of green, varying from light bright green to a deeper hue: some of the larger, older, and fruit-bearing fronds being almost of a bluish-green. The extensively creeping root-stocks are black, and fibrous, sending up fronds from buds at the spreading angles of the roots ; the young fronds are clothed with thin, white, loose scales, which are easily dispersed. The fruit dots are at first a pale whitish-green, but deepen to brown, and in some old strong fronds are almost black. This last form may possibly rank as a variety ; it certainly differs in size and colour : it has larger, stouter, darker fronds ; the stipe not less than a foot from the base to the lower divisions of the leafy portion of the frond ; the long slender stipe is greener, and the whole plant coarser and more spreading ; the pinnules more oblong and bluntly crenate at the edges.

It is the smaller fronds that are so very attractive from their bright vivid colour, dark stems and triangular form. This fern is found in woods among mossy roots of old decaying stumps and rotten wood, among which it sends out its slender running root-stocks and loose black fibrous rootlets. I should think this beautiful fern would be rather difficult of house or pot culture, unless its peculiar soil could be provided, which would be difficult to introduce on a limited scale, but it would grow well out of doors in a wilderness or grove, especially if indulged with a rotten log or stump.

### Shield Ferns—Wood Ferns.

This interesting and hardy family contains many distinct species with numerous varieties, some of which have as yet hardly been recognized by the Pteridologist. The Aspidiums are mostly confined to the forest or swamp ; some seem to rejoice in the deep shade and soil of the thick Pine-woods, while others are to be found in wet spongy soil as Beaver meadows, Cedar swamps, and boggy, spongy soil among tangling weeds and bushes. In such situations we shall find the

### Marsh Fern—*Aspidium Thelypteris*, (Swz.)

It is especially a moisture-loving Fern, and may be seen forming thick beds in low lying ground, which is often over-flowed, in open marshy spots, especially where shallow springs gush out among grass and sedges ; in such places it is sure to be found sending up its light green, rather broadly lanceolate, fronds, which are stiff and upright ; the stipe, dark coloured, often twisted, but not bending ; the pinnæ horizontal, but slightly bending downwards ; pinnules a little revolute at the edges ; root-stock creeping, sending up the fronds from buds at intervals ; rootlets black, very fibrous ; the fertile fronds are twice as long as the sterile ; stipe nearly double the length of the leafy portion of the frond ; pinnæ narrowly pointed, distant on the rachis ; pinnules narrowly contracted ; sori forming a line partly protected by the revolute edges, terminating the upper forking veinlets ; many fine, white, shining glandular hairs may be seen at the revolute edge of the pinnules, which, in the early stage of the fruit dots, nearly cover them.

A very different appearance is presented by those plants which grow in the dense shelter of Cedar swamps ; the sterile fronds are lax, almost drooping, of a dark green colour ; pinnules blunter, distinct ; stalks of a very dark colour ; pinnules farther apart ; fertile frond from three to four feet high ; stipes very nearly three times the length of the frond, narrowly contracted, dull and dark ; root-stock extensively creeping ; the pinnæ drooping from the apex of the frond, sometimes again curving upwards.

Another form of *A. Thelypteris* is found in swampy spots, the pinnæ wider, very much deflexed and thinner, light green ; stipe slender, green. This fern is found on Long Island in Rice Lake, it was named for me as a form of *Aspidium Thelypteris ;* it is rather fragrant. I have never been able to procure one of the fronds in a fruiting state. The lax drooping pinnæ, and oblong blunt pinnules, with their pleasant sweet scent, distinguish this fern from the common forms of the Marsh Fern.

NEW YORK FERN.—*Aspidium Noveboracense* (Swz.)

Is closely allied to *A. Thelypteris*, but is a decidedly more elegant Fern. It is narrowly oblong, lanceolate, pale green ; the lower pairs of pinnæ small and deflexed, very scantily developed ; the upper pairs very sharp and pointed, giving a narrow pointed outline to the fern, which, in the delicate young fronds, is peculiarly graceful. The fronds grow from a circular crown and bend outwards ; the root-stock is prolonged and slender, creeping and fibrous, throwing up the clusters of fronds at intervals. The fertile fronds are generally about as tall as the sterile but are rigid, and the pinnules narrowly contracted, opposite ; sori of a deep rich coffee-brown, becoming confluent so as to cover the underside of the frond ; indusia kidney-shaped. The soil where I found an extensive bed of these very graceful, plumy ferns, was rather sandy, at the edge of a wood known as Preston's Wood, not far from the Village of Lakefield. The pinnules of this fern are narrowly oblong, sometimes serrate at the edges. In the month of October the fronds fade to a delicate buff, when the sori are distinctly seen on the pale ground forming a distinct border.

This fern is pleasantly fragrant when drawn through the hand or slightly bruised. In colour, it is yellowish-green where growing in

light soil and exposed to the sun, but becomes darkened in the shade
and more lax in habit ; it is a remarkably elegant gracefully growing
species, very narrow in outline, and finely diminished to a point at the
apex of the frond, and is a valuable addition to the fern-garden.

EVERGREEN WOOD FERN.—*Aspidium spinulosum*, (Swz.) var.
*intermedium*, (D. C. Eaton).

This handsome evergreen fern forms one of the greatest ornaments
of our Canadian woods, where it exists in great abundance, and in
several various forms.    Not only is it one of the earliest, but one of the
hardiest and most enduring of its tribe.    We find its refreshing verdant
fronds, as soon as the snow wreaths have disappeared from the forest,
brighter and greener than any other herb or tree that has borne the
pelting of the pitiless storms of Winter.

Our Wood-fern affects the soil of the Pine and Hemlock woods,
in which many other species of plants refuse to grow ; but it is also
abundant in hardwood lands, where, however, we find it more usually
growing at the foot of Pines or Hemlocks, as most congenial to its
nature.

The large, green feathery fronds are from one to two feet in the
full grown plant, sometimes exceeding this measure in favourable
situations, and varying in colour from a lighter to a deeper green.  The
stipe is rough, channelled, stiff and scaly, both in the bud and up the
stalk, but the scales are less abundant in some plants than in others.
The fruit dots are round, kidney-shaped : at first pale, but deepening to
dark brown, almost black, and shining.    This last appearance seems
peculiar to very large spreading vigorous fronds, which may be another
variety of the typical form of *A. spinulosum.*    The pinnules much
divided almost tri-pinnate ;  wide-spreading, upright, and of a deeper
green than the more common form, which is narrower ; the pinnæ
curving upwards, the fruit dots covered with a pale brown indusium.

A more delicate narrower fern, with pointed scales, shaded with
darker brown in the centre, not more than a foot in height, the pinnæ
elegantly curved upward, fruit dots smaller, of a very light brown, may
be the *Lastrea collina* of Moore.    There are other not less attractive
varieties, which may be found in our Pine forests, intermediate between
the Evergreen Fern and *Aspidium dilatatum*, which also has its own
varieties uniting it to *A. cristatum.*    Many of these gradations among
our Aspidiums are yet un-named, unless by such a fanciful name as seems
appropriate from some peculiarity in outline or colour.    One of the
most beautiful of these varieties we might call the Lace-Fern, from the
graceful appearance of the deeply divided pinnules, thinness of texture

and bright green colour of the frond. One form of Lady Fern very narrow in outline, pinnæ very distant, and pinnules almost folded together, is known as Skeleton Fern, and this last name is very well suited to the starved aspect of the plant. While one of the Beech Ferns* has obtained the singular name of Mendicant Fern, from the deflexed lower pair of pinnæ, which have a sort of supplicating look. The elegant *Adiantum pedatum*, is known everywhere in this district as Fairy Fern† ; nor can we reasonably object to common names by which our plants may be recognized by common people, who would only provoke a smile were they to call them by their scientific names, the meaning of which they could not comprehend nor even pronounce properly.

The fruit dots of the Evergreen Wood Fern may be perceived on the back of the pinnules as soon as the frond unrolls ; at this early period they have a whitish look, and the kidney form is easily distinguished. In July the colour deepens, the indusia shrivel and the ripened sori give a rich brown look to the back of the frond ; in August the spores begin to be shed abroad.

What a world of wonders does the magnifying glass reveal to our eyes if we examine the fruit dots through it. Truly those who never look within the book of Nature, lose a thousand pleasures that they never dream of, in their eager pursuit after worldly amusements. Even the fine transparent pointed hairs that terminate the toothed divisions of the pinnules are most beautiful to look upon ; the fine veinings and the scales that clothe the root and stipe of the frond, are worth our closest attention and admiration. Very closely allied to *A. spinulosum* var. *intermedium* is the

BROAD SHIELD FERN—*Aspidium spinulosum*, (Swz.), var. *dilatatum*,
(Horneman).

This differs from the preceding by its more triangular outline, the oblong divisions of the pinnæ, the more leathery texture and yellower green of the pinnules and the paler, smaller, fruit dots, which are very abundant on the full-sized fruitful fronds. Where this fern is found, in exposed situations, the colour is a very yellow green, rusty on the stipes, and the pinnules contracted, so as to appear concave beneath. The rachis is thickened and a little swollen at the base of the mid-ribs of the pinnæ. These differences evidently are caused by soil, and more sunny exposure, from which all our wood-loving ferns seem instinctively to shrink, nor do they long continue in situations so uncongenial to their

---

* *Phegopteris hexagonoptera.*
† These are merely local names not widely known.

nature, for they dwindle away and disappear after the second year  Like the var. *intermedium* it seems to prefer the Pine-woods, and is also found very frequently in Cedar swamps among rotting wood; it then assumes a bright glossy green, and has the pinnules larger and more dilated.

This very handsome, hardy fern dries well, especially the young fronds, which retain their bright colour and are extremely attractive. The sori on the fruit-bearing fronds, do not (or very rarely) extend to the two lowest pairs of pinnæ. Like other ferns of this tribe, the frond is circinate in the bud, the root-stock very scaly, and the rootlets strong, black, and fibrous.

MARGINAL SHIELD FERN—*Aspidium marginale*, (Swz.)

Of this fine hardy fern we have three distinct varieties besides the more common normal form, known as the Marginal Shield Fern, and described by Gray as of a light-green colour. This is not the colour of our Canadian fern which is decidedly dark, or very full green.

The fern described in the Manual agrees perfectly with fine specimens found growing in a swampy, rocky, bush road, near the banks of the river Otonabee, about a mile south-east of the village of Lakefield, but differs from the more common dark-leaved evergreen fern of the Pine-woods in the same neighbourhood, which is of a fine shining holly-green colour, densely chaffy at the root-stock; stipe, and rachis green, and smooth, about two to three feet high in the larger fronds which are pinnate; pinnules crenate; veins, waved; sori, abundant, marginal, leaden coloured, but turning brown when fully matured; indusia, depressed in the centre, kidney-form as in other Aspidiums. This is a stately evergreen fern, common in rich damp woods, especially where Pine and Hemlocks abound in rocky woodlands.s

A very interesting small sized fern resembling the above, only of a miniature growth, I have found in the rocky woods near the Otonabee; and also on a bit of wettish waste-land two miles north of Peterboro', among rotten logs and limestone boulders. Stipe slender, smooth, very chaffy at its root-stock, from four to ten inches in height. Many of the fronds were fruitful at six inches in height, but usually the fruiting fronds were the largest and strongest. The veinings were forked, and the mid-veins blackish and waved; sori, distinct and formed at the upper fork of the veinlets; colour of the indusia, at first whitish then of a leaden hue; the scales at the junction of the stipes with the root-stock of a pale fawn-colour, soft and thin pointed, and over-lapping as in the typical form. The favourite soil appears to be decaying wood, as it was chiefly found growing on decayed logs, between the chinks or sheltered on the ground below. The pinnules slightly crenate, oblong, blunt, and the whole plant of a dark glossy-green.

### Mrs. Traill's Shield Fern—*A. marginale*, (Swz.) var. *Traillæ*, (Lawson).

A more remarkable form of *A. marginale* was found near the village of Lakefield, on a vacant town-lot, still only partially cleared from the forest trees and brushwood. Rearing its noble dark-green fronds among the broken piled up branches of a brush-heap, I found the fine tall fronds of the dark-green fern to which Professor Lawson has given the name *Traillæ*, in compliment to the finder.

From a hard, woody, chaffy root-stock, standing some inches above the soil, close to the roots of an old Beech stump, sprung up some six stout fronds of a deep-green colour; pinnæ long and narrowly pointed, curving upwards; pinnules deeply crenate so as to form wide sinuses between the lobes; pinnæ distinct on the rachis nearly to the upper pairs; indusia pale-green, kidney-form, one on each lobe, for with the exception of one or two smaller fronds, they were all fruit-bearing; veining strongly marked, twice-forked, and waved, so as to form an elegant sort of shell-like pattern on the upper surface of the leaves. The height of this stately fern was from two to four feet. The root-stock was evidently many years old. The only second plant of this fern that I have since found was at the side of a piece of corduroy road at the foot of a hill near Mr. G. Strickland's, but it was in a mutilated condition, having been bitten by cattle; nevertheless it was not so much injured but that it could be readily identified with the former specimens, which fact was very satisfactory. These plants grew nearly two miles apart, but fire has destroyed the one, and the other has disappeared beneath a newly erected fence, to my sorrow and disgust.

This fern is so distinct in its features that I think it may be considered a species rather than a mere variety. Prof. Lawson's description is as follows: " Fronds very large (3½ feet long), bipinnate, all the pinnules pinnatified." He further states that this variety has the same relation to the type of *A. Marginale* as var. *incisa* has to typical *Filix-mas*.

### Crested Shield-Fern.—*Aspidium cristatum*, (Swz.)

We meet with the Crested Shield Fern in great abundance growing in the Cedar swamp in the rear of the village of Lakefield, North Douro. This very handsome, though somewhat coarse fern, grows freely in wet boggy ground, among fallen timbers and rank herbage; often on rotten logs or at the roots of stumps. The young bright green fronds are extremely pretty. Sometimes the roundish lobes of the pinnæ are crowded and stand forward on the rachis, so as to give a full crisp look to the

Q

plant. I have also found an abnormal form of *cristatum*, where the pinnæ were upright, clasping the rachis and twisting round it ; the pinnæ were very narrow, as also the lobes ; the sori flatter and the indusia much thinner and paler in colour ; the fronds of a very light green, and the lower part of the stipe of a reddish brown.

I have also met with a chance form where the pinnæ were forked at the lower pairs, but accidental varieties occur in most of the forest ferns.

I have also found specimens of *Cystopteris bulbifera* and of *Dicksonia pilosiuscula*, handsomely forked at the apex.

The fertile frond of *A. cristatum* is much taller than the sterile frond. The stipe is of a rich reddish brown colour ; a few large, thin, pale-coloured scales may be observed on the stipe and rachis. The pinnæ are broad at the base, slightly petioled, near the lower part triangular, the lobes divided within a little of the mid-rib, toothed, and tipped with sharp points, whence the specific name *cristatum* is doubtless derived. In growth this fern is upright, rigid and stiff, but when in full maturity has a very rich appearance, from the abundance of ripe sori. These at first are pale-green, then they deepen to a dull leaden colour, and finally are rich brown, abundant, but not often confluent. The barren fronds are usually of a darker green than the fertile, which become of a light yellowish colour towards August, and finally are rusty and discoloured. The sori are ripe in August and September.

### Larger Shield Fern.—*Aspidium Goldianum*, (Hook.)

This is nearly allied to *A. cristatum*, but is a much larger species, more triangular in the arrangement of the long curved pinnæ, of a darker fuller green, less spiny and chaffy ; the blunt pinnules are slightly toothed : the large brown fruit dots are arranged in a distinct row (not confluent), nearer to the mid-veins than to the margin. The pinnæ are long, sometimes in fine fronds from four to six inches from the rachis to the extremities, while the height of the larger fronds will be from two to four feet, where the soil is rich and damp. Like all the *Aspidieæ*, this fern is circinate in the bud ; the frond for the ensuing season being green and ready to be unrolled early in June, or in warm Springs late in May, protected by a hard scaly covering from the inclemency of the Winter's cold.

The pinnules on the divisions of a fine robust specimen of *Aspidium Goldianum*, now before me, are twenty-eight in number, each of the largest pinnules three-quarters of an inch from the mid-rib to the blunt extreme end, while there are no less than thirty-eight pairs of pinnæ, exclusive of the last few, which become confluent at the apex of the frond. Though the stipe of this fern is very stout and upright, it is easily broken and yields to the influence of weather, and later in the season may be found bent and bruised, splitting into long strawy threads. The stipe is of a pale colour, smooth and shining, not channelled as in others of the tribe, where the strong nerves are more apparent.

This fern seems, like *A. cristatum,* to grow most luxuriantly in damp woods, Cedar swamps, and beside creeks in thickets.

The young fronds of *A. Goldianum* are beautifully veined ; veins and veinlets dark and finely waved.

On the whole, though coarse, it is a grand and stately fern, especially when the sori are perfectly ripe. It is a hardy species, but the fronds wither down in the Autumn.

HOLLY FERN.—PRICKLY SHIELD FERN.—*Aspidium (Polystichum) Lonchitis,* (Swz.)

This is a somewhat coarse, robust fern of stiff upright figure, from a densely, hard, chaffy root-stock. The young fronds lie closely rolled up during the winter months, protected by the brown scaly covering, a characteristic mark common to all the hardy species of this order of ferns—the Aspidieæ—many of which are of an evergreen habit, remaining healthy and bright through the storms and snows of our Canadian winters. The name Holly-Fern no doubt is derived from its dark green colour and the rough spiny appearance of the fronds. The whole plant is clothed from root to apex with pale, golden-brown pointed scales and hairs, even the veins and mid-ribs are thus clothed in old strong-growing plants. The stipe is short, thick, channelled, and clothed with coarse broadly-ovate scales. The divisions of the pinnæ are auricled or lobed on the upper side, somewhat hollowed or curved towards the centre, and pointed at the apex of each one, sharply serrated and finely bristle-tipped; veins free, bearing the round, rather small fruit dots in a regular line, not quite at the margin of the pinnule, and the fruit-bearing pinnæ seem to be confined to the upper portion of the frond. The pinnæ on the lower part of the frond diminish in size and extend nearly down to the root-stock. Though not one of the most elegant of our native ferns, the Holly-Fern is yet ornamental, from its fine dark-green colour, evergreen habit and hardy nature, which makes it eligible for out-door culture in artificial rock-work. The habitat of the Holly Fern is in the crevices of rocks, and it is found abundantly at Owen Sound, in the same locality as *Scolopendrium vulgare.* It is not so tall growing a plant as the

CHRISTMAS FERN.—EVERGREEN ROCK FERN.—*Aspidium (Polystichum)*
*acrostichoides*, (Swz.)

This handsome dark glossy fern is abundant in our northern townships ; it is of a full deep shining green, lanceolate in outline, the height
varying in some of the old plants from one to two feet ; the upper portion
of the fruiting frond is contracted, and closely covered with the confluent
fruit dots ; the whole of these narrowed pinnæ have a fine brown felted
appearance from the abundance of the indusia.    The edges of the pinnæ are strongly revolute, which gives a hard and rigid look to the frond ;
but these contracted pinnæ only occupy about a third of the leaf.    They
are long scythe-shaped, auricled, or very destinctly lobed, the margins
sometimes toothed and tipped with fine silvery hairs.    But there is a
great diversity in this fine evergreen Polystichum ; the edges in the
younger barren fronds are often quite smooth, while in others they are
bluntly notched, or again finely serrated, and fringed with shining hairs ;
before unrolling the fronds are densely clothed with white silvery scales,
which give a soft woolly look to the rolled up frond ; but the white
scales disappear very soon, or take a browner hue as the plant increases.
I consider *A. acrostichoides* is much finer as an ornamental fern than
*A. Lonchitis*; this last is much coarser in texture, the pinnules are short and
sharply cut at the edges, and grow along the whole length of the rachis
from root to apex, the sori are small, round, dark and do not contract
the edge as in *A. acrostichoides*: the whole plant is more upright, stiffer,
and wanting the rich, smooth, glossy surface of the Evergreen Rock Fern
of our woodlands.    There are several handsome species of the genus
enumerated by the British botanist, but they do not appear to have our
handsome Christmas Fern in England.

HAIRY WOODSIA.— *Woodsia Ilvensis,* (R. Br.)

We must not seek for this pretty fern in our rich Woodlands, it is
a rock-loving plant, and chiefly found in the black-friable soil that lies in
the crevices of rocks, into which its black fibrous roots can easily
penetrate.  It may also be seen in grassy places near water, but always in
rocky localities.

From a thick clump of matted black roots arise a number of rather
slender stipes, terminated by narrow lanceolate fronds of a leathery
texture, and deep green colour ; the pinnules are closely sessile to the
mid-rib, the pinnæ also closely adhering to the rachis ; the fruit dots
are abundant on the slender tapering fertile fronds, covering the obtuse,
about three-lobed, pinnules with the pale brown hairy indusia, which are
early disrupted, and form thin pointed scales, surrounding the fruit dot,
almost resembling the persistent calyx of a flowering plant.

In length the tallest fronds of this pretty fern do not exceed twelve inches; the barren and younger growth of fronds vary from a few inches to six or seven. The fronds are circinate in vernation, being closely rolled in, and when first breaking the ground the whole of the frond is covered thickly with white silvery hairs, which assume a fine light, shining brown after exposure to the light and air, as the season advances.

Very fine specimens of the Hairy Woodsia were sent to me by the kindness of Mr. Watt, from the neighbourhood of Montreal, and also by my much-esteemed botanical friend, Professor Macoun. Though not uncommon in some localities near Montreal and eastward, it was unknown to our forest settlement at that time; but it is found among the rocks that form the western and southern barrier, that surround that little gem-like lake, known by the settlers in Smith and Burleigh as Fairy Lake, a little tarn of some acres in extent, separated from Stoney Lake by a wall of rocks. This lonely spot is known chiefly to the hunter and trapper, but latterly its loneliness has been invaded by pleasure seekers, and sought out where it lies,—

" 'Mid circling rocks tha hide it from the world."

To some such, it may appear rugged and savage in its wildness of rocks and trees, and tangling underwoods; but not so to the Botanist and Field-Naturalist, who will find in it many flowering shrubs and ferns, such as the noble *Osmunda regalis* and the small but interesting *Woodsia*, one of the most charming of our Rock-ferns, and well worth the attention of those persons who delight in artificial rock-work in their gardens or pleasure grounds. One would think it would be easy of culture, as it seems hardy in its natural condition, where, with Alpine hardihood, it braves the cold of our Canadian Winters and rocky heights unhurt. Many of our ferns once little known or valued, have been sought out and brought into notice by the labours of the intelligent members that form the Botanical branch of the Field-Naturalists' Club of Ottawa. All praise be due to the men who have made known, by their industrious researches, the riches that have been so long unnoticed in our forests, our rocks and our waters.

To a certain class of minds these things appear trivial and of no value; they do not see that the power of a nation does not consist only in trade and what arises from its commerce alone, but in the intelligence of its people, and in the natural productions of the soil, which, being sought out and made known, are—through the mechanical skill and inventive genius of others—the source of a nation's wealth and greatness.

GOSSAMER FERN—*Dicksonia pilosiuscula*, (Willd.)

Gossamer Fern, a charmingly appropriate name for one of the most beautiful and graceful of our native ferns, more delicate in the fine texture of its frond than the Fairy Fern (Canadian Maiden-hair) or the slender Cystopteris. From the delicacy of the finely divided pinnules of the pinnæ, which terminate in points so fragile that a breath disturbs them, it is very difficult to preserve this lovely fern in perfection when drying, and its symmetry is destroyed unless great care is taken not to ruffle or move the tender thing till it has been some days under pressure and has become sufficiently dry to change the drying sheets of paper.

The Gossamer Fern is the latest of our wood ferns to expand and unroll, the upper end of the fronds remaining curled up until the month. of June, unless the season be warm and moist. The stipe or stalk is very stout, hard, and at first of a bright rosy tint, especially in the young growth ; in the older fronds the colour becomes of a rich golden brown, smooth and glossy. The scales of Dicksonia are few, round or blunt, thin, and pale in colour. There is a fine, minute pubescence on the surface of the very light green pinnules. The pinnate divisions of the frond are set close upon the rachis, crowded, opposite or alternate. I have numbered from forty to sixty pairs in one frond ; the pinnules are deeply divided to the mid-rib, finely toothed. The fruit dots are small and round, placed at the point of the full forking veins in cup-like involucres. The outline of the fronds is pyramidal, broad, and wide below, and narrowing to the most delicate point at the extreme apex and also of the pinnæ. The root-stock is slender, running below the surface and sending up many fronds at intervals, forming extensive beds where the soil is light and the ground shaded.

It was in a wood known as Preston's woods, Lakefield, that I met with the beautiful Gossamer Fern growing in two large beds which had been divided by a road passing through them.

A more lovely sight than these beds of light green feathery ferns presented to the eye I had never seen. So fair, so fragile they looked, contrasted with the dark sombre shade of the great rough-barked Hemlock trees beneath which they grew ; their delicate forms, so fairy-like, just stirred by the least breath of summer wind, gave one the idea of the utmost frailty ; yet these slight fronds bore the effects of the early nipping frost more hardily than the coarse Bracken and some of the stouter herbs around them.

Later in October I found the delicate green of the *Dicksonia* changed to a fine buff or creamy-white colour, shewing the ripened fruit dots distinctly on the light, faded frond, as beautiful in that faded state as in its summer verdure.

The Gossamer Fern appears to increase more by the extension of its creeping root-stock than by seed. From the strength and stoutness of the older fronds bearing the sori, I should think that this plant requires the growth of several successive years before producing them. The root-stocks of many of the ferns I believe to be of considerable age, having known the same plant year after year growing on the same spot.

The vernation of *Dicksonia* is different from that of the Aspidiums, the fronds being encased in sharply pointed buds, not rolled in the same way, and covered by the brown chaffy scales that protect the hardy wood ferns from the frosts of winter.

SENSITIVE FERN—OAK-LEAVED FERN—*Onoclea sensibilis*, (L).

This fern abounds in low wet spots, marshy flats, and on the low-lying banks of lakes and slow-flowing streams, where there is shade and the soil is suitable ; it grows to the height of three feet and upwards ; much depends upon the age and strength of the thick, black root-stock, which extends horizontally, sending up stiff upright stipes, and large triangular fronds, which vary considerably at different periods of growth, and are circinate or rolled when they issue from the bud ; being at an early stage, covered with fine white velvety down, which afterwards disappears.

The seedlings of the Oak-leaved Fern, *Onoclea sensibilis* differ with succeeding years ; at first a tiny, delicate, little club-shaped, three-lobed plant, semi-transparent of a very light-green colour ; the next year it has made a great start in size and darkness of colour. We next find it a strong vigorous fern with the divisions of the frond, long, waved at the margin, and having veinings distinctly marked ; rhizome, thickened, extending under the surface with strong, black, wiry roots and buds ready for the next year's growth ; for the next few years the leafy fronds increase in size and substance, but still make no attempt at forming fertile fronds. When full grown it is no longer like the original seedling that we noticed among the damp, spongy, peat soil of the low swampy river bank ; but a large robust plant spreading abroad its coarse, wide-expanded frond. After a time, from the crown of the rhizome rise some shorter fronds, having smooth stipes, and instead of being foliaceous the pinnules are short, rolled up and converted into berry-like involucres, forming a one-sided panicle.

The sori are round and borne on the back of the veins, and are quite concealed within the berry-like involucres.

Later in the season the normal fertile fronds turn dark-brown, and remain persistent till the following year.

The age of the root-stock of this fern appears very considerable, indeed we know not how long it endures, sending up by extension new fronds, year after year. Where it grows in shady spots, by river and lake shores, the growth is rank, the texture coarse, and the colour a very dark sombre-green. The sterile fronds are sometimes very large, occasionally attaining to the length of two, or even three feet ; the horizontal root-stock extends a considerable distance ; this last may be said to have been the cradle of the undeveloped fronds, for years past. By dissecting the rhizome the embryonic green frond can plainly be seen. Not only can the process of preparation for the coming year's growth be seen in the ferns, but the same thing is observable in the leaf-buds of the deciduous trees and shrubs, where, as soon as the leaf begins to decay, a new one is forming, ready to burst forth from its winter casing in the ensuing season.

In the occasional form found growing in low ground, which is called var. *obtusilobata,* and which has the pinnules extended as in other ferns, instead of being rolled up in the usual berry-like manner, and where the sori are much less abundantly produced, it can be seen that each sorus is protected by a very thin hood-like indusium fixed at its lower side.

This form occurs in large beds of the normal fern not unfrequently, but appears to be accidental. It is probably an imperfect state of the fertile frond. Several very pretty specimens of another abnormal state of *Onoclea sensibilis* occurred on the shore of Rice Lake. These fronds were of small size, not more than six to eight inches in height, and were distinctly bi-pinnate ; the rounded, blunt pinnules were crenate, and each pinnule petioled. The veins were straight and then forked, and being of a deeper colour than the leaves, gave a beautiful variegated look to the frond. This was an intermediate state between the normal form and the so-called var. *obtusilobata.* I have since met with several specimens of similar appearance, and one or two of these shewed a few fruit dots on the margin of the upper divisions, which has induced me to conclude that by some accidental cause the perfect fruition of the frond had been prevented.

It is very gratifying to an amateur Botanist to imagine that he or she has found a new species, or even been the means of establishing the identity of a doubtful one ; but it is more to the advantage of the science of Botany that one truth should be established, than many new species added to the list of those already known. I think that the variations in the appearance of *Onoclea sensibilis,* arising from the increasing age of the plant, have led some persons to conclude that there are two or more varieties. It would be presumptuous in one, with such limited knowledge as I possess, to assert that there were no other permanent forms of this fern than one common one, but I know that every year makes a difference in the outline and texture of the frond that might easily deceive a casual observer.

The fertile, or fruit-bearing frond, like that of *Onoclea Struthiopteris*, is distinct from the sterile or foliaceous one. It does not appear till late in the month of August or the beginning of September, and the seeds do not ripen till October. It remains persistent through the winter, and old stalks will continue through the following summer, till, battered and worn, they decay and disappear. These fertile fronds seem to be the production of the root-stock of several years' growth, certainly not less than three, and probably the perfect condition may not be arrived at under four or five years' growth. I have noticed large beds of Onocleas continue year after year without a single fruit-bearing frond, while in beds of older and stronger plants the seed-bearing plants were abundant.

The appearance of the fertile frond of Onoclea is very handsome and remarkable ; the stiff, upright, rigid stipe is surmounted by rows of round berry-like spore-cases, formed by the contraction and altered condition of a foliaceous frond. This may be observed both in the very early and immature states of the fertile frond, and later when the spores are shed and the ragged, worn envelope shows whence it had its origin. At its first appearing the colour of the seed-vessels is of a bluish green, but as the season advances they become of a dark chestnut-brown, and have a pretty round bead-like appearance, forming a close, upright, compound spike, each branch being formed by a pinus, and each berry-like process by an altered lobe. Perhaps we have not a more interesting fern among our Canadian species than *Onoclea sensibilis*, as a study, from the tiny little club-shaped seedling to the old and perfect plant.

OSTRICH-FEATHER FERN—*Onoclea Struthiopteris*, (Hoffm.)

In wet, marshy ground, on the swampy flats of low lying lakes and boggy meadows, or on the banks of shaded creeks, we meet with the fan-shaped, coarse, but grand looking Ostrich-Feather Fern. The large, strong fronds rising from a central caudex form a circular crown slightly bent outwards, making a graceful plume-like figure, whence its popular name has been derived, and which is well suited to the picturesque arrangement of the fronds.

This fern is recommended as being easy of growth, bearing removal from its native soil even when the fronds are well developed, provided the pot, or box, or place in the rock-work be kept well-watered and somewhat shaded from the sun. There is a great difference between the plants growing in open and exposed places, and such as are under the shade of forest trees, the latter being more luxuriant, deeper coloured, and attaining to a far handsomer, more plume-like form ; the fertile fronds are very dissimilar from the sterile. They are the production of plants of several years' growth from old, woody, root-stocks, on short, stout, deeply-channelled, strongly-nerved stipes ; the fruit dots are round, on the free veins and veinlets of the contracted pinnules, which are rolled back, forming a rigid bead-like object, the two opposite lines meeting together on either side of the mid-rib, and forming a covering to the numerous clusters of sporangia, which by their thickened substance they effectually guard and conceal. At first these singular fruiting fronds are green, but assume in maturity a rich coffee-brown and take the semblance of the quill-feathers of a dark plumed bird. So perfect is the deception, that at first sight you marvel how these stiff brown feathers got stuck into the heart of the tall, graceful, waving circle of Ostrich-Feather fronds that surrounds them. These fertile fronds remain persistent all through the Winter and late into the Spring of the following year, but the winds and rains of Autumn and the frosts and snows of Winter begin to wear the surface and tear the cover that was so tightly secured, and then you may detect the leafy substance and nature of the pinnules, and perceive the veinings of the leaves ; and the secrets that were so carefully concealed, within the now torn and ragged outer coating, are laid open to the curious eye of the Naturalist.

The stipe and rachis of the Ostrich-Feather Fern are, as I observed before, deeply channelled, and the sides rendered convex by two stout elastic nerves on either side of the stem. Near the root-stock the stipe is flattened like the handle of a spoon, but tapers to a narrow point at its insertion with the root-stock. The flattened part is smooth, slightly hollowed, black, and polished like whalebone, and finely fringed with short, stiff, glandular hairs.

The fronds for the ensuing season are circinate in vernation, early shewing the hard, round, rolled up contents beneath the thick covering of pale, membranous, chaffy scales ; the roots are strong, black and wiry. The root-stock attains to a great age, and becomes very hard, black, and woody.

BRITTLE BLADDER-FERN—*Cystopteris fragilis,* (Bernh).

This truly graceful fern affects the chinks and crevices of limestone rocks, under the overhanging shelter of bushes and long grass, where it is shaded from the ardent rays of our July and August suns. It was in the limestone rocks in the quarry at Lakefield, near the Otonabee rapids, that I first discovered tufts of this charming fern. There, nourished by the moist atmosphere from the river, almost hidden from sight, it adorned the rude rock with its drooping slender fronds of tender green. The stipe is about a third the length of the frond, of a dark chestnut colour at the lower part, smooth, shining and lax, in young fronds almost thread-like. The rhizome is tufted, throwing out an abundance of drooping fronds from small buds collected into a crown. The rootlets are black, delicate, and numerous ; rhizome running, but not extending to any length. The brown sori are numerous, but not often confluent, situated at the base of each veinlet which terminates the toothing divisions of the blunt and somewhat rounded pinnules. The lower pinnæ are smaller than the next three or four pairs, these last are distant, curving upwards, becoming closer towards the summit of the frond, but scarcely confluent on the rachis. On a fine specimen now before me there are eighteen pairs of pinnæ ; the whole length of this elegant frond, from the root to the apex, measures eighteen inches, the brown, shining stipe being about a third of the length of the foliaceous portion.

In the variety *angustata*, the frond is more upright, of a darker green ; from the narrowness of the pinnæ and pinnules, the fruit dots are closer, and form an almost continuous line, of a rich, dark brown colour ; the closeness of the sharply toothed divisions gives an elegant pale green fringed edging to the pinnules, the effect of which is very graceful. The covering of the fruit dots soon breaks away, and the seeds are ripe early in July. The extreme delicacy of texture in *C. fragilis*, causes it to wither earlier than most of our ferns, especially if the summer be very hot and dry ; it generally disappears early in August. The seedling plants are very delicate, lovely little things. The root-stock does not appear to increase much till after the third year, producing no fruiting fronds till after that time. Thus, at least, I have concluded, after close observation of the plants for several years past.

There are many variations in the form of this fern, some fronds being broader at the base, the pinnæ closer set, and fruit dots more distant, the stipe greener and shorter, and the toothing divisions more sharply cut ; but I think it is merely a variation which may be found in many plants rising from the same root-stock.

I believe that a number of specimens lately gathered in the limestone rocks on the Smith side of Lakefield may be referable to *C. dentata* ; but as the plant presents distinctive characters I shall describe it more particularly :

Root-stock tufted, fibrous ; fronds circinate ; stipe and rachis, bright, reddish-brown, smooth and slender, the stalk rather waved than drooping ; colour bright light-green ; pinnæ horizontal, opposite ; pinnules bluntish, toothed, but not so much divided as in *C. fragilis*, consequently the fruit dots are more distant, the veinlets being less forked.　The hood-like indusia are of a pale colour; the stipe much shorter than the frond, which is long and lanceolate; the pinnæ are distinct to the last pairs, which are very minute.　One of the most remarkable features in this fern is the presence of bulblets, (which I have never seen on the true *C. fragilis.*)　These bulblets are of a deep green colour—not thrice cleft and mitre form as in *C. bulbifera,* but more like the swelled lobes of a bean—one being nearly twice as large and overlapping, the other.　These bulblets are partially covered with broadish, thin white scales, which probably in an early stage enveloped them, but now appear disrupted and scale-like.

In *C. bulbifera* the thrice cleft bulblets stand upright, but in my newly found fern they lie horizontally on the upper pinnules.

The root is blackish, fibrous, wiry, imbedded in the crevices of the limestone rock, above the river.　The whole plant is from six to nine inches high.　I think that it is nearer to *C. bulbifera* than *C. fragilis.* It is altogether a very graceful and interesting little plant, and would be very suitable for rock or parlour cultivation.

The larger fern, *C. bulbifera,* with the variety *flagelliformis,* grows in vast quantities in our damp woods, where it trails upon the ground or climbs over fallen timber and brush-wood, sometimes stretching its weak attenuated stalk to the length of three and even four feet, and in several specimens I have detected delicate white fibrous rootlets put forth at the extreme end of the frond, as if to clasp the bark or lay hold of the surface of the stump, or branch, that was nearest, for support.

This fern might be called the Trailing Fern, from its lax habit. When young it is broader at the basal pinnæ, of a vivid green, but becomes much lighter, almost yellowish-green, in the fertile state of the frond.　The stalk is red, smooth and brittle : the few loose scales that appear early in the season soon drop off and disappear; the lower pinnæ are longer, wider at the base, and curve upwards, they are distant from the next pair, and so continue till about the third or fourth pair, when they are more horizontal on the rachis and opposite, continuing distinct, however, to the end.　The sori are whitish at first, but become of a dark brown colour as the indusia shrivel up.　In one form, it may be the variety named by Dr. Lawson, *horizontalis,* the fertile frond is almost folded together, the pinnules being contracted so as to show but one side.　There is a form of very frequent occurrence, the whole frond being stiffer, more leathery in texture, and of a duller colour ; the fruit dots very dark brown.　Further observation may throw more light upon this plant to determine whether this appearance is constant or accidental.　If distinct, it might be called variety *rigidus* from its stiff rigid habit.　The continual changes that are being effected on the face of the country, render accurate attention to the peculiar habits of our native plants daily and yearly more difficult. To-day I go forth into the woods and discover some interesting plant, which I desire to see unfolded in perfection ; a few weeks pass, and lo ! the axe of the chopper has done its work, the trees are levelled to the earth, the fire has passed over the ground and the blackness of desolation has taken the place of verdure and living vegetation. I must seek my plant in more distant localities, and it may be it is lost to me for ever, and I console my disappointment by the consideration that such things are among the "must be" of colonial life, and so it is useless to grumble.

The Greek name *Cystopteris*, is derived from two words, signifying —a bladder, and—a fern, from the inflated indusia or hood-like coverings of the seed.

ROYAL FLOWERING FERN—*Osmunda regalis*, (L.)

" Fair ferns and flowers, and chiefly that tall fern
So stately, of Queen Osmunda named."—*Wordsworth.*

The name Osmunda, in the Anglo-Saxon tongue, means "Peace" or " House-peace "; so says Moore, an English writer on Ferns. A sweet feminine name, worthy of being borne by some gentle maiden in her rural home, as well as bestowed upon the noblest and most attractive of the Fern Family, or, as it has been styled by a modern author, the " flower-crowned prince of ferns."

The old and long-recorded legend of the origin of the name Osmunda, given to this fern, is simply that Osmund, a prince, some say a waterman—possibly both (for the prince might have owned a boat, in which he amused himself on the waters of Loch Tyne)—is said to have secreted his wife, or a fair daughter, during an incursion of the savage Danes, among the tall fronds of these ferns, whence the name was given by the people in remembrance of the safe shelter afforded by the tall, shadowing fronds to the fair " House-peace " or the lady Osmunda. Now the story may only be a pretty fanciful romance of Anglo Saxon days, but we like to indulge in a bit of poetical fiction when we can, in regard to the names of our pet plants ; there is something in the name and its attributes—Royal Flowering Osmunda— surely it sounds better than such Latinized or Greekified names as we find in our Botanical Catalogues. What of such names attached to some of our Sedges as Carex Shortiana, C. Hitchcockiana, C. Sartwellii, C. Schweinitzii, C. Muskingumenis, C. Wormskioldiana, and a host of others; but I am on dangerous ground, or I shall get into disgrace with my Botanical friends, who will hardly forgive my impertinence and presumption, or ingratitude to the dear professor, who named a doubtful fern after myself.   But to return to my Osmunda ; the root-stock of the *Osmunda regalis* is very hard, stout and knotty in age, often standing many inches above the surface of the ground and sending up many slender stipes, which are circinate in the bud, but not so warmly clothed with chaffy scales as the Aspidiums and others.   The stipes of the young, and also of the sterile fronds, are of a rosy red tint, deepening into blackish brown near the root in the old and fertile fronds.   I know not a more beautiful appearance than a grove of these stately ferns makes in their bright, lively green, summer dress, above which is seen the rich cinnamon brown tufts of the fruit-bearing pinnæ, bending their graceful foliage over the brink of some lonely lake as if to kiss the reflection mirrored on its surface.   The pinnated divisions of Osmunda are opposite, or nearly so ; the pinnules of the older, stouter fronds are an inch in length and half an inch in width, not very close on the mid-rib : those of the younger and sterile fronds slender and paler in colour.   There is a great variety in the foliage of this fern ; sometimes the pinnules are eared or auricled, the margins waved, notched, crenate, or finely serrate (saw-toothed), or in young ones quite smooth, small and delicately thin : the stipes red, slender and drooping gracefully downwards.

The indusia appear to be lobes or parts of the pinnules transformed into berry-shaped masses that conceal and cover the sori, and open with a slit, when the seeds are ripe, to let them escape from their carefully sealed prison.   What wondrous care, what consummate wisdom is here displayed by the Creator in the protection of the life-containing germ of a simple fern.   It is as if a sort of maternal instinct had been imparted to the parent plant to shield the embryo from every possible injury and to insure its safety through all the mysteries of its infant state, till the time should come for it to be launched forth to find a home and nourishment in the bosom of the earth.   Our fern is, by some authors, supposed to be a distinct variety from the British, but as the description given in Moore's " Handbook of British Ferns " describes our's exactly, I cannot do better than reproduce it :

" The fronds are circinate in vernation, and when young, delicate and very tender, shooting up with rapidity and attaining in some places a height of ten or twelve feet in damp ground, in drier situations from two to four ; they appear in May and are destroyed by early frosts. The stipe is stout, smooth, without scales, tinged with red while young ; fronds, lanceolate, bi-pinnate ; pinnæ arranged in opposite pairs with opposite or alternate pinnules, often lobed or auricled at the base, serrated or otherwise ; the venation very distinct ; each pinnule has a prominent mid-vein ; the veins from this are forked, and the venules are again forked in parallel lines to the margin. The fructification consists of the upper pinnæ, changed from a leafy to a soriferous state, and forming a panicle of spikelets, covered over with spore cases attached to the veins of the altered pinnules.

The spore cases are sub-globose, reticulated, two valved, opening vertically, these valves being supposed to originate in the epidermis of the frond."

In its faded state, when touched by the withering power of Autumnal frost, the Osmunda is not less beautiful in its warm buff-colour, and with its darkened plumy crests consisting of the ripened panicle of sporangia which are borne aloft on the summits of the main stems. I have seen a grove of Osmundas towering above that remarkable plant, the *Sarracenia purpurea*, with its dark crimson-veined hollow cup and ewer-shaped lip. The Pitcher Plant itself, springing from the midst of a deep bed of creamy peat-mosses, the myriads of soft leaves and rose-tinted capsules of which were alone a sight to charm the eye of the lover of nature, not less lovely because untouched by artificial culture, and fresh in all their native grace and beauty, adorning the waste places of the earth, wild and free from God's hand.*

WATER FERN—INTERRUPTED FERN—*Osmunda Claytoniana* (L.)

Water Fern is the name by which this fern is known among country people, from its being indicative of hidden springs. Where these ferns are found, even on high ground, there is every chance of water existing below the surface, so the old settlers used to say : a more satisfactory way of discovering springs than the far-famed mystery of the Wych-hazel wand of the well-diggers which many persons put implicit faith in.

The old name *Osmunda interrupta*, or Interrupted Fern, given by Michaux and older Botanists, is so expressive of the peculiar character of this remarkable plant, that I like to preserve it, as it has a distinctive reference to the arrangement of the pinnæ, which is peculiar to this fern, and by which it can be recognized by any observer of the plant.

The fronds, when they first appear above the ground in May, are densely clothed with light-brown woolly hairs ; as they unroll, the fertile contracted pinnæ are seen, like a cluster of brownish-green caterpillars ranged in pairs, from three to four or five, occupying the middle of the

---

* At the right hand corner of Fairy Lake, in a piece of White Peat Moss, *Sphagnum cymbifolium*, already described, this fern towered up far higher than the head of the writer, and above that of the boatman, H. Stone, who was a man above the middle height.
At the side of a water-course near Rice Lake, fronds of *O. Claytoniana*, above the height of the lady who gathered them, were brought to me, and she was five feet six inches.
*Woodwardia Virginica* is also a tall fern ; on a rocky ravine at the north side of Eagle Mount, on the Dumones side of Stoney Lake, there is a large thicket of this fern, in which were growing many fronds, which exceeded in measurement the stature of my companion, who was a person very little under six feet. These examples show that the above statement of the height of the *Osmunda regalis* was no exaggeration, though possibly above the average.

rachis, or rather nearer to the base ; the lower pairs of sterile pinnæ below these curious rolled up ones, are smaller than the upper leafy pairs which expand into a fine ovate foliaceous frond.

There is an interruption in the outline of the frond, caused by the closely contracted fertile pinnæ. At a first glance you would really imagine that you saw a number of hairy caterpillars feasting on the green frond. As the season advances, these narrowed contracted pinnæ acquire a deep brown colour, still bearing a strange resemblance to some predaceous insect ; by the end of July they begin to shed the spores and shrivel up, but cling for a while to the rachis till they finally drop off, leaving a naked vacant place on the stalk between the upper and lower leafy portions of the frond. The soft, woolly covering that we first noticed has almost disappeared ; it had fulfilled its part in the economy of the plant : like a warm and comfortable great coat, it had guarded the young plant from injury during the capricious season, when it made its first appearance, and now is cast off as no longer needed. The usual colour of our Osmunda, during its earlier stages of growth, is a very light yellowish-green, but as the fronds increase in size and the summer advances, the now largely developed fronds (and truly it is a stately plant) become of a deep, sad green. I have measured full-grown fronds of *Osmunda Claytoniana* five feet six inches in height, and I have been told that many are found still higher.

The usual habitat of this fern is the dry beds of old water-courses, swampy places and wet meadows. The stipe and rachis are yellowish in old plants. The root-stock survives to a great age ; I knew a plant growing in one spot just below the orchard at Oaklands, Rice Lake, my old home, for more than forty years ; when last I visited the spot the same old root was still sending up its annual cluster of noble fronds, as fresh and as grand looking as when first I noticed the plant, in the year 1840.

The age of some species of ferns, especially that of the Osmundas and the Onocleas, appears almost to reach that of a tree, for, having paid much attention to these interesting plants, I have known the same old standard root-stocks for a number of years past, unchanged in their character, unless it were that they became more vigorous in sending up larger and more fruiting fronds.

The subject seems to have excited little attention hitherto, but is a fertile field for investigation, which I recommend to my readers.

In olden times the early settlers made use of the young, tender heads of this fern as a pot-herb, likening its flavor to that of Asparagus. One point in its favour may be noticed, that the vegetable is quite harmless and may be used without fear of its being poisonous.

PLATE IX.

I. LARGE EVENING-PRIMROSE (*Œnothera biennis* var. *granaiflora*).

I. TWIN FLOWER (*Linnæa borealis*).

### CINNAMON FERN—*Osmunda cinnamomea*, (L.)

This fern is so called from the reddish-brown wool that clothes the fronds in their early stage of growth, and also from the colour of the sori of the fertile frond which occupies the central crown of the root-stock, quite different, and easily distinguished by its reddish colour and contracted pinnæ, from the upright, coarse, foliage of the sterile fronds.

The Cinnamon Fern is found in wet, grassy places, growing in large clumps, sometimes from very large, hard, fleshy root-stocks, which are occasionally eaten raw and are said to resemble some nuts in flavour The singular-looking red-brown fertile frond rising from the midst of the plant, soon perfects its seeds, and the long weak stipes fall and lie prostrate on the ground, curled up and withered. Though the sterile fronds are coarse and less elegant than those of any of the Osmundas the very young leaves are remarkably handsome. The pinnules which are roundish and blunt are usually entire and crowded, nearly over-lapping on the short chaffy stipes, the free forked veins easily discerned.

### ADDER'S TONGUE—*Ophioglossum vulgatum*, (L.)

The Adder's Tongue in general appearance hardly realizes the idea of a fern, although in reality it is one. Closely allied to the Adder's Tongue, and belonging to the same Natural Order, are the Moonworts. This order which Botanists name *Ophioglossaceæ*, includes those ferns whose leaves instead of being rolled are folded up in a straight or inclined manner. The sporangia are formed of the interior tissue of the frond, and are spiked or panicled. The copious, sulphur-coloured spores are discharged through a transverse slit which divides the spore case into two valves. The roots are fleshy and instead of sending up several fronds produce one double frond every year, which is divided into a leafy barren portion, and a fruitful portion, which is simply a spike or cluster of sporangia. The Adder's Tongue is not very common in Canada, it is found in open spots, among grasses, near woods. The leafy portion of the frond is simply an undivided, entire, egg-shaped, fleshy frond of a light green, and is beautifully veined. The fertile spike is borne above this expansion on a slender stem, and consists of two rows of spore cases, one on each side of the stem, forming a double row about an inch long. When the spores are ripe these cases split open and have the appearance of a double row of teeth; it is from this state of the plant that the name Adder's Tongue, which is a translation of the classical name, is taken. The height of this strange little fern is about six inches. There is generally only one frond from the same root; but occasionally two may be found.

R

FLOWERING WOOD-FERNS—RATTLESNAKE FERNS—GRAPE FERNS.

The Botrychiums differ from the Adder's Tongue in having the barren portion of the frond divided into a series of stalked pinnules, and have the fertile portion made up of alternate clusters of sporangia. The Moonworts are pretty plants, with much more of the appearance of ferns than the Adder's Tongue ; but possessing many distinctive characteristics which distinguish them from the Dorsal Ferns—as the Polypodies, Aspleniums and Aspidiums—for instance, the forked fronds dividing the stipe into two distinct parts : the sterile or foliaceous, and the fertile or fruit-bearing portions ; the smooth, fleshy stipe destitute of scales ; the berry-like, rolled up sporangia that contain the spores, and the thick, fleshy roots, covered with a tough, leathery, brown skin, which, in the older plants, is marked with indented rings, and from which, doubtless, the name of Rattlesnake Fern has been derived. Another distinctive mark is the collar or crown from which the bundle of thick roots springs ; the central bud contains the single foliaceous frond, which is not circinate or rolled up, as in the Dorsal Ferns, but is folded in a thin, white, membranous sheath, similar to those that enclose the flowers of bulbous-rooted plants. The sporangia, the coverings of the spores or fern seeds, are on an upright slender stipe, much longer than the sterile frond, which is sessile, or closely adhering to the main stem. These spore cases are formed of altered pinnules, which are contracted and closely rolled up into tiny balls, opening, by means of a slit, into two valves, which allows the fine yellow or pale brown seeds or spores to escape when ripe. Of this interesting family we have several distinct species and several varieties. In Britain, I believe, the only representative of the Botrychiums is *B. Lunaria*, the Moonwort, a small, singular looking fern, with semicircular, thickish, crenate pinnules and slender fertile spike. Respecting this curious plant many incredible legends existed in the days when that worthy old herbalist Gerarde wrote, but which he disclaimed as not proven.

The English peasant, though far from imaginative, is credulous when once impressed, and readily accepts the marvellous, only it must have the sanction of ancient custom and oral tradition, handed down from father to son ; this sanctifies and gives weight to any legendary lore, however improbable. It must have been a hard trial when purer faith replaced the legends of the church with simple Gospel truths. *Botry-chium Lunaria*, (Swz.) has been lately found in several localities in the Provinces of Quebec and Ontario, but is very rare. The commonest species is

### Rattlesnake Fern—*Botrychium Virginianum,* (Swz.)

This fine fern may be found in the rich vegetable mould of our hardwood forests, and can be easily distinguished by its broad bright green, much divided, barren frond of thin texture, with the veins free and forking; pinnules slender and pointed; the upright, fertile frond much taller, slenderly branched, bearing the small round seed vessels of a dark green colour, ripening to a bright reddish-brown or yellow, and shedding the spores in the month of July, or early in August. In some specimens the leafy frond may be found placed obliquely, giving a more graceful air to the fern.

The variety known as *B. gracile* I take to be plants of the first and second years' growth. Having cultivated this fern for many years in a wild shady spot in my garden, I have watched it during its different stages from the first seedlings as they came up, just simple miniatures of the larger growth, at first small, with leafy spreading surface, cut and divided as in the older plants, but not forking into two divisions. The second season these plants became fruitful : though the fertile spike was very delicate, the whole size of the fern not exceeding six inches ; as it increased in age it became stouter in the scape and larger in breadth, corresponding to the description of Gray. My observations were not confined to one or two specimens, but I grew and examined a large number, and was convinced that I had formerly been mistaken in supposing that there were two or more distinct varieties, age only being the cause of the difference. I had also thought that this fern was one of annual or biennial growth, but this also was a mistake, as I have learned by experience and observation that the root-stock is of longer duration.

In very damp, rich soil, *B. Virginianum* attains to a large size—from one to two feet in height, and from twelve inches and upwards in breadth.

The thick, brown, fleshy roots strike far down into the soil, but though difficult to dig up uninjured, the plant is not hard to cultivate, as, with water and shade, in good mould, it grows and thrives well in the garden, and the seeds spring up early in the following Spring—though it seems to me only a very few of the thousands of fern seeds shed possess the life principle within them ; possibly a wise provision in nature to keep these plants within proper bounds, has restricted them from vegetating freely.

The name *Botrychium* is derived from the Greek *botrys,* a bunch of grapes, from the clustering of the spore cases.

## Moonwort—The True Rattlesnake Fern—*Botrychium ternatum,* (Swz.)

This is a very distinct species from *B. Virginianum,* differing in colour texture and shape of the pinnæ and in the general outline of the plant. The thick dark-green colour and fleshy consistency of the frond, with the blunt eared or lobed form of the pinnules, the thicker midribs of the sterile, divided frond, and tall fruit-bearing portion, two to four pinnate, presents a very different aspect to that of the above mentioned fern. The old year's frond remains, persistent and evergreen, until the new one makes its appearance, breaking its way through the sod from its enveloping sheath. I found this fern in grassy Pine groves, just appearing above the ground in the month of July, on the banks of the Katchawanook Lake.

One of the forms of this species was formerly called *B. lunarioides* (or lunaria-like) on account of a resemblance it is supposed to bear to *B. Lunaria*—but it, with several other forms, is now included in the type of the species *B. ternatum* (Swz.)

The light brown fleshy roots of this fern are tough and deeply ringed, whence its name Rattlesnake Fern no doubt has been derived.

Another very striking form of this species I discovered in the Oaken glades of the Rice Lake Plains ; it differed in some particulars from the above. In height, not exceeding nine inches ; the stipe very thick and stout, very smooth, and when wounded emitted a thick juice which gave a white starchy crystal when dry. The last year's leaf remained, sheathing the stem and falling prostrate as soon as the new frond appeared, and began to expand its thick leaflets and forking fertile frond. The colour of the whole plant was of a light yellowish-green ; the sporangia yellow ; the spore dust sulphur yellow, shedding abundantly when ripe, which was in the hot month of August. The outline of the fronds was nearly circular ; the fertile frond very little higher than that of the leafy or sterile portion, which is closely sessile to the column-like fleshy scape, and spreads in a fan-like manner.

As the house in which I lived was on a sloping bank, above the valley where found my ferns, I had a good opportunity of observing their peculiar habits and progress from day to day ; and marking the difference from those that grew in the Oak brush on the open plain-lands and the plants that grew among the Pine scrub. The soil of the Oak-lands was sandy, or light loam, while that on the banks of the Katchawanook was gravelly ; the difference of soil might account for the difference in the growth and colour of the foliage, the one being of a deep sad green, the other very light, of a yellowish tint ; the one tall, the other very low and stout. I think it was a distinct variety from

Gray's figure of *B. lunarioides.* The mucilaginous juices of this fern would lead one to suppose that it might be of a softening and healing nature. The British plant *B. Lunaria* was reputed to possess very healing qualities, and I can readily believe that our native Botrychiums, from the soft mucilaginous juices of the stem when broken, would be found softening and healing when applied to wounds or tumours.

# CANADIAN FERNS

—SUITABLE FOR—

## Cultivation in the Old Country.

———

I T frequently happens that Canadians are applied to, by their friends in the Old Country, for roots of our native ferns for cultivation in Ferneries and Conservatories, and it is not infrequently the case that these requests are not complied with, either from a lack of knowledge as to what species would be acceptable, or from the mistaken notion that, because a fern is exceedingly abundant here it must also be so in Great Britain. As a guide to those who may be anxious to delight the hearts of their Old Country friends, by sending them living roots of Canadian ferns, I append lists —First—Of those which are not found growing indigenously in Great Britain at all ; and Secondly—Of those which, although found there, are very rare, or which present differences in appearance from our forms, in consequence of which they would be acceptable to all Fern-growers and Collectors for comparison.

In the cultivation of ferns there are one or two points which must always be borne in mind. First, it must be remembered that ferns, unlike most other plants, shun the sun-light and court the shade and a moist atmosphere. Although some species may be occasionally found growing in open spots, it will generally be found that they are more luxuriant when in the shade and under the shelter of trees. Many ferns will succeed well in flower-pots ; but they require constant care and attention. One of the most important requirements of ferns is perfect drainage ; and this can only be attained at the risk of allowing the roots to become too dry, unless regularly watched. A liberal supply of drainage, in the shape of broken pots, pieces of old mortar, or bricks, should be placed in the bottom of the pot, and on the top of this a thin layer of moss, to prevent the fine mould, which is necessary, from running down into the drainage. The mould should be light, and finely sifted. A useful mixture for most ferns is the following : Fibrous Peat, Leaf-mould, Loam and White Sand, in equal proportions. The roots bear transplanting better in the Autumn than at any other time, but should not be forced to grow at once on arrival in Britain ; they should be put by in a cool place until the next Spring.

The facilities for the transmission of small parcels to Britain are now so great, that there is no excuse for our not sending large numbers of these lovely plants, many of which, although common with us, are, nevertheless, highly prized on the other side of the Atlantic.

## Ferns Indigenous to Canada, East of the Rocky Mountains but not found Growing Wild in Great Britain.

### FAIRY FERN—*Adiantum pedatum*, (L.)

An exceedingly desirable species, of great beauty and easy culture ; there is, perhaps, no species of the large family of Maidenhairs, from all the different parts of the world, which surpasses our native species in grace and elegance.

### ROCK BRAKE—*Pellæa gracilis*, (Hook.)

This charming little species has yellow, fleshy root-stocks, which contain much oil, and it is doubtless this character which renders it capable of withstanding drought for a long time. Its natural habitat is in crevices and under overhanging ledges of limestone rocks, where it shoots out its delicate fronds in the month of June. At this time of the year the rocks in such localities as it frequents—on river banks and lake shores—are constantly wet from the snow water, which has not yet dried up in the woods ; but is constantly trickling down over the surface of the rocks, and which, penetrating into the little cracks and crannies, gives the moisture which is necessary, with the warmth of Spring, to quicken into life the lovely rock ferns which grow there. After the end of July the fronds dry up and the plant again lies dormant until the next June. This fern could be moved easily when in the dormant state, and should be cultivated with ease.

### CLIFF BRAKE—*Pellæa atropurpurea,* (Link)

A very attractive plant of small size, found in the crevices of dry rocks and cliffs ; it should transplant easily, and, with care, would succeed in artificial rockwork where there was not too much moisture.

### CHAIN FERN—*Woodwardia Virginica,* (Sm.)

This would perhaps hardly succeed so well in cultivation as many of our other ferns, as it is a vigorous-growing, coarse plant, with large and long root-stocks, which grows in peat-bogs and Tamarac swamps. Its fine appearance, however, makes it worthy of a trial.

### EBONY SPLEENWORT—*Asplenium ebeneum,* (Ait.)

Great care must be taken in the removal of this and all the Spleen_ worts of the same class from their native rocks. These rock-loving Spleenworts will, however, generally succeed well if their roots are taken up intact, and if they are planted in crevices of rock-work, or even in flower-pots if well drained. In rock-work they should have an upper position, but should not be kept too dry. They generally succeed better if planted in a horizontal manner between two stones ; when grown in pots, the soil should be a light sandy loam mixed with leaf-mould, and among this, about the roots, should be placed some pieces of old mortar or sandstone.

### NARROW LEAVED SPLEENWORT—*Asplenium angustifolium* (Michx.)

This is an exceedingly handsome fern, well suited for conservatories. It requires good rich leaf-mould or peat and plenty of moisture, when the large, delicate, light-green, fronds would be produced in abundance and form a charming contrast with the darker hue of other ferns. The name of this fern is perhaps a little deceiving. Although narrow-leaved it is one of our largest Spleenworts, the fertile fronds frequently exceeding two feet in length.

### SILVERY SPLEENWORT—*Asplenium thelypteroides* (Michx.)

This is another of the large-fronded moisture-loving Spleenworts. Its natural habitat is in the deep forest or in swampy woods. It is well fitted for conservatory and greenhouse culture, and like the preceding, should have deep leaf-mould and plenty of moisture. The fronds are very handsome, of a deep green, and, as the name implies, the plant has much the appearance of the Marsh Fern (*Aspidium Thelypteris*) and grows much in the same manner, throwing up clumps or clusters of tall graceful fronds.

### WALKING FERNS—*Camptosorus rhizophyllus* (Link.)

This is a particularly acceptable species, found on shaded limestone rocks. It is an Evergreen, is easy to cultivate and has a very distinct appearance, quite different from any British fern. It will transplant at any time of the year, and if given a shady corner with good leaf-mould, plenty of moisture, and sufficient room to spread out its fronds, will grow luxuriantly, and increase rapidly by taking root and making young plants at the tips of the fronds.

### MENDICANT FERN—BROAD BEECH FERN—*Phegopteris hexagonoptera* (Fee.)

This resembles the British Beech Fern (*Phegopteris polypodioides*) very closely, but is a larger and handsomer plant with fronds of a delicate light green. It is easily cultivated, but must have good deep leaf-mould and shade.

### CHRISTMAS FERN—*Aspidium acrostichoides* (Swz.)

This is a dark handsome fern, of a deep full green. A very desirable species, hardy and easy of culture, and has a very effective appearance among other more delicate ferns. It is always to be obtained without difficulty in rocky woods, growing in clumps or small beds. It has somewhat the appearance of the Holly Fern, but is more showy.

### NEW YORK FERN—*Aspidium Noveboracense* (Swz.)

A delicate and attractive species, not difficult of culture. Should have light soil and plenty of moisture ; but must be well drained. This fern is always acceptable, both from the delicacy of its foliage and its soft green colour.

### GOLDIE'S SHIELD FERN—*Aspidium Goldianum*, (Hook.)

A large, rich and handsome species, with dark foliage, that has lighter shades down the centres of the pinnæ, which give it a pretty, variegated appearance. It is easy of culture, and well suited for the back of a rockery. It succeeds better when supplied with a liberal allowance of peat and leaf mould, but is very hardy, and will grow in almost any soil, with shade and moisture ; it becomes stunted, and seldom produces fertile fronds unless provided with good soil, moisture and shade. There is no fern in Britain that has the same appearance as this ; and this fact makes it a desirable species.

MARGINAL SHIELD FERN—*Aspidium marginale*, (Swz.)

A handsome fern, but bearing too close a general resemblance to *A. Filix-mas* to make it a desirable species for transmission to Britain. *A. Filix-mas*, although so rare on this continent, like *Scolopendrium vulgare*, is one of the most abundant species all over Great Britain.

EVERGREEN WOOD FERN—*Aspidium spinulosum*, (Swz.) var. *intermedium*, (Gray).

This common variety of *A. spinulosum* which is found everywhere in our woods, is not among those which grow in England. It is the *A. Americanum* of Davenport, and is a graceful plant, which should be included in all collections of ferns sent to Europe. It is hardy, very easy of culture, and easily obtained. The fronds are evergreen and of a bright colour.

BOOTT'S WOOD FERN—*Aspidium spinulosum*, (Swz.) var. *Boottii* (Gray).

This is a much rarer fern than the preceding, with thicker and less cut up foliage, sometimes resembling *A. cristatum* as much as *A. spinulosum* in appearance, and much more like the British form of *A. cristatum* than the Canadian; it is considered, however, to be a variety of *A. spinulosum*. It requires a good depth of leaf-mould, shade, and a liberal supply of moisture.

CLINTON'S CRESTED SHIELD FERN—*Aspidium cristatum*, (Swz.) var. *Clintonianum* (Gray.)

This handsome fern is quite unlike any European form of *A. cristatum*, and in some respects bears a much closer resemblance to our own *A. Goldianum*. Its habitat is wet swampy woods and in the deep forest, where it sometimes grows to a large size. It is a very desirable species for European collectors to compare with their own forms of *A. cristatum*.

SCENTED SHIELD FERN—*Aspidium fragrans*, (Swz.)

There are few of our native ferns more attractive than this; the deep blue-green fronds hang in rich clusters from the crevices of rocks where it grows. It is considered one of our rare species, but is generally to be found in abundance where it occurs. The Lake Superior region seems to be the centre of its distribution; there it is most abundant, growing on trap rocks. The close compact growth of the plant and the abundance of the sori with their lead-coloured indusia, give this fern a very rich appearance, and its agreeable scent makes it still more a desirable species. It would probably be rather shy of cultivation, but might be treated in the same manner as the Rock Spleenworts.

### TRAILING FERN—*Cystopteris bulbifera*, (Bernh.)

No collection of Canadian Ferns would be complete without this common, but charming species. It is undoubtedly one of our most elegant ferns; the slender, elongated fronds of light-green, with their ruddy, semi-transparent stipes, render it most valuable for contrast. It is very easy of culture, and will flourish luxuriantly if planted on a wet wall, or near a waterfall. It grows easily in a flower-pot in a conservatory, but cannot bear sunlight. It should have a light peaty soil, and plenty of moisture, both in the atmosphere and at its roots. It would probably grow better if small pieces of rock were placed among and about its roots.

### OAK-LEAVED FERN—*Onoclea sensibilis*, (L.)

There are few of our ferns which are greater favourites in Europe than this common species. It has long been known there as a greenhouse plant, where it is prized as well for its graceful foliage as for the ease with which it can be cultivated. With light soil, shade and moisture it grows well in Europe and increases rapidly, but seldom produces the fertile fronds so abundant here. It is always acceptable.

### OSTRICH-FEATHER FERN—*Onoclea Struthiopteris*, (L.)

This stately plant is valuable for the back of a rockery. If supplied with a good depth of leaf mould it grows easily and throws up its lofty plumes to the height of three or four feet or even more. There is no fern which resembles it in Great Britain. It bears transplantation well, and will grow vigourously in a flower-pot in the house, when it becomes a useful and graceful ornament for a drawing room.

### NORTHERN WOODSIA—*Woodsia hyperborea*, (R. Br.)

All Woodsias are welcome additions to a European collection. Unluckily, however, all of our species with the exception of *W. Ilvensis* are very rare. *W. hyperborea* is only found on high cliffs and near waterfalls. The culture should be the same as for the Rock Spleenworts, the chief essential being perfect drainage, together with plenty of moisture and light soil.

### SMOOTH WOODSIA—*Woodsia glabella*, (R. Br.)

This species is found in similar localities with the above, and should be cultivated under the same conditions.

## Gossamer Fern—*Dicksonia pilosiuscula*, (Willd.)

This lovely scented fern, with its delicate fronds of tender green, is a charming object as it grows in its native woods, where it is generally found on a cold, sandy loam. It is, however, rather shy of cultivation, The root-stocks are very wide-spreading, and in transplanting specimens, small and young plants should always be taken. A sandy loam, with about one-fourth leaf-mould, is the best soil for this fern. Although difficult to grow, its beauty well repays any trouble expended upon it.

## Interrupted Fern—*Osmunda Claytoniana*, (L.)

This fern and the next are always welcome additions to European collections, not only for their own beauty, but because they are so different from the Royal Flowering Fern, *Osmunda regalis*, which is well known there. The Interrupted Fern is well suited for planting in fountains and on the borders of ornamental ponds, where its curious, graceful fronds are shown off to great advantage.

## Cinnamon Fern—*Osmunda cinnamomea* (L.)

This is another handsome fern, generally found in slightly wetter places than the preceding, but it is hardier and may be grown under the same circumstances. The foliage is not so handsome, but it is always an acceptable plant and grows easily.

## Rattlesnake Fern—*Botrychium Virginianum* (Swz.)

This is a great favourite with British collectors. It is very different from the European *B. Lunaria*, grows easily, is of convenient size and compact growth. Light soil, about half leaf-mould, is the best for this fern. It succeeds well, when single plants are grown separately in flower pots·

## Grape Fern—*Botrychium simplex* (Hitch.)

This rare fern has more the appearance of *B. Lunaria* than any of our American representatives of the genus. It, like *Botrychium matricariæfolium* and *B. lanceolatum*, is difficult to cultivate, and the only way to succeed with these species is to remove a large quantity of the soil with them from their native haunts so as not to disturb the roots. When growing in turf in meadows this is not difficult, but when in the light leaf-mould of the forest it is not so easy

## Ternate Moonwort—*Botrychium ternatum* (Swz.)

This is a very handsome, dark green, fleshy fern, found in open grassy spaces near woodlands. It has thick fleshy roots, but is rather difficult to move unless a piece of sod is taken up with the plant. The fertile frond is large and conspicuous, and is a great ornament to the plant when the spores are mature. It is an Evergreen, and takes a rich bronze tinge of colour from the winter's frosts.

---

## Ferns Indigenous to Great Britain and Canada but Rare in in the Former country, or which preŝent such Differences in Appearance as to make them desirable for Comparison.

### COMMON BRAKE—*Pteris aquilina* (L.)

The different aspect this plant bears in Canada, to the tall, luxuriant fern of England, attracts the attention of all visitors from the Old Country. The chief differences consist of the smaller fronds and much heavier and more conspicuous fructification. The two are considered to be identical but it would be interesting to grow them both together and observe whether the differences remained constant.

### LADY FERN—*Asplenium Filix-fœmina* (Bernh.)

Under this name are grouped many widely separated varieties, all of which would be interesting for comparison with the British forms. The beauty of all the varieties of this fern, however common they might be, would always render them acceptable to fern growers.

### BEECH FERN—*Phegopteris polypodioides* (Fee.)

This fern which grows so abundantly in most of our woods, is much less common in Great Britain, where it is also known under the name of the Mountain Polypody. Our plant differs somewhat in appearance and habits from its British representative. The habitat of the Canadian plant, in Ontario, is on rather dry knolls in woods, and the fronds are almost of a leathery consistency: while in England, the Mountain Polypody is found in deep shade on the margins of mountain streams or in damp woods ; the fronds too, are of a much more delicate texture and colour than in our plant.

### OAK FERN—*Phegopteris Dryopteris* (Fee.)

This lovely, delicate species, which is found everywhere in our woods, is quite uncommon in England ; it is more abundant in

Scotland, but is very rare in Ireland. It will always be acceptable when sent to Britain, and is of very easy culture, throwing out in every direction its slender, creeping, scaly, root-stocks, which bear a profusion of lovely yellowish-green fronds. It requires plenty of moisture and shade, but will grow readily in pots in the conservatory or in the open air. The frond is distinctly divided into three divisions, from which it is sometimes called the Ternate Polypody. This trifoliate character of the frond is distinctly seen in the young fronds before they unroll, when the three divisions appear as three little balls on slender stems at the summit of the stipe. This distinguishes it easily from the

### LIMESTONE POLYPODY—*Phegopteris calcarea*, (Fee.)

This is an exceedingly rare species in Canada, having been found only once by Prof. Macoun, on the islands of Anticosti in 1883, and by Messrs. Bell and Dawson at the Lake of the Woods. It is a much less rare fern in England than here.

### MARSH FERN—*Aspidium Thelypteris*, (Swz.)

It may seem strange to Canadians that this fern should be included in the list of desirable species to be sent to Europe ; because, being found there at all, it might naturally be supposed that its rapid growth and hardiness, which make it so abundant everywhere here in low, swampy ground, would also have the same effect in Great Britain. This, however, is not the case ; and although pretty generally distributed over England, it is by no means common, while in Wales and Ireland it is rare, and in Scotland very rare, only one county so far having produced it. It can be grown with the greatest ease in pots, which may even stand in water. It is a most graceful object in cultivation, the lovely, delicate, almost semi-transparent sterile fronds being produced in the greatest profusion. In this, as in some others of our wild ferns, when cultivated in the conservatory, the fertile fronds are seldom produced.

### CRESTED SHIELD FERN—*Aspidium cristatum*, (Swz.)

This plant, although appearing on the British lists of ferns, is exceedingly rare, and should by all means be included in all collections of Canadian ferns sent to Europe, not only because of its relative rarity there, and abundance in this country, but because our fern bears only a very slight resemblance to the British fern which goes under the same name. Our plant has long, narrow fronds, of a thick, leathery, almost coriaceous texture, while the British plant bears a closer resemblance to some of the forms of *A. spinulosum*, and, strange as it may appear to us here, where these two species have aspects so dissimilar, in England botanists are sometimes at a loss to know to which of the three species, *A. cristatum, spinulosum* or *dilatatum*, certain specimens should be referred.

Thomas Moore, the author of a magnificent volume, entitled " The Ferns of Great Britain and Ireland, nature printed " even going so far as to say " Indeed, so closely do these merge into each other by means of transition forms of frond, that we are forced to the conclusion that they are all three in reality mere variations from one specific type."

What we call *A. cristatum*, is an abundant species in low wet woods and swamps, easily obtained, and grows readily, the root-stock throwing out lateral extensions not found in var. *dilatatum*.

### Wood Fern—*Aspidium spinulosum*, (Swz.)

This species is so variable, that it, with its many varieties, would alone furnish sufficient material for a special study. We have several forms in our woods more or less permanent. They would all be interesting to compare with the English forms. One of the handsomest varieties, var. *dilatatum*, presents characters which would almost claim for it the rank of a species. The most important difference is that the root-stock of var. *dilatatum* is large and tufted, and the crown is often raised somewhat above the surface of the soil, forming a short caudex, while in *A. spinulosum* the root-stock has a creeping habit, so as to become branched or multiplied in time into several tufts and clusters with crowns, as in *A. cristatum*.

This is an easy fern to cultivate. It requires a good depth of leaf mould or peat, plenty of moisture, and if favoured with a shady nook will produce fronds of great beauty.

### Brittle Bladder Fern—*Cystopteris fragilis*, (Bernh.)

This is another fern which will always be acceptable to British collectors. It is not at all common in Europe as compared with its great abundance in North America. It is an exceedingly variable species, so much so that the name may almost be said to cover a group rather than a single species. It is a very desirable species for greenhouse or out-door cultivation, growing rapidly and with little care. Being a rock fern, it should have pieces of old mortar or limestone placed in the mould about its roots, and will grow luxuriantly if supplied with plenty of water, and at the same time kept well drained. It is easily multiplied by division of the root.

### Oblong Woodsia—*Woodsia Ilvensis* (R. Br.)

This fern, which is abundant in Canada, being found from the Atlantic to the Rocky Mountains, is exceedingly rare in Great Britain and for this reason is particularly acceptable to collectors. It seems to be rather shy of cultivation, and curiously enough, seems to flourish best when taken least care of. On our Canadian rockeries it can be grown easily, and probably would succeed better in England in the open air than in the conservatory. It should have an upper position in the rockery where it gets complete drainage. It is a plant of compact growth and great beauty.

### FLOWERING FERN—*Osmunda regalis*, (L.)

The great beauty of this common species would always make it acceptable, even in parts of England where it grows. Nowhere, however, is it so abundant as it is here, in our low meadows and swamps. Our plant, too, has somewhat a different aspect from the British, being smaller, and more ruddy in the colour of the young fronds. The Canadian form has been called *O. spectabilis* by some botanists; but we occasionally find large specimens in Canada which cannot be distinguished from the British form.

This is a moisture loving fern, and must have a good depth of leaf mould. It makes a grand ornament for the back of a rockwork, and will produce gigantic fronds if planted in a large pot and placed in the basin of a fountain or on the edge of an ornamental pond.

END.

S

# INDEX

Printed in the United States
By Bookmasters